Environmental Monitoring with AI and IoT

Environmental Monitoring with AI and IoT: Smart Systems for a Sustainable Future aims to provide a comprehensive understanding of how modern technologies like Artificial Intelligence (AI) and the Internet of Things (IoT) revolutionize environmental monitoring. It bridges the gap between traditional environmental assessment methods and advanced, technology-driven approaches to enhance efficiency, accuracy, and scalability.

This book emphasizes the practical application of AI and IoT in addressing pressing environmental challenges, such as air, water, and soil quality monitoring, while aligning with sustainability goals. The scope of this book is vast, covering both foundational concepts and advanced methodologies in environmental monitoring. It includes essential topics like sensor technologies, AI techniques for data analysis, IoT communication protocols, and ethical considerations in environmental monitoring. Real-world case studies and practical examples make the content relatable and actionable.

Students, instructors, and professionals will benefit from the insights into creating and deploying smart environmental systems capable of addressing diverse challenges, from urban air quality management to disaster prediction and response. Students will find the content highly accessible and aligned with academic and industrial requirements, while instructors will appreciate its structured approach to teaching these cutting-edge topics.

Sneha Gautam is an Associate Professor in the Division of Civil Engineering at Karunya Institute of Technology and Sciences, Coimbatore, India. He has received numerous awards, including the Young Scientist Award 2023. He also serves as an Editor, Associate Editor, Editorial Board Member, and Guest Editor for several high-impact journals. With over 146 articles in leading journals, he has developed a global network of collaborators, and his research has been featured in prominent media outlets.

Chafai Azri is a prominent researcher specializing in environmental science, with a focus on air quality, pollution, and ecological health risks. His work has significantly contributed to understanding the dynamics of atmospheric pollutants and their implications for human health, particularly in the context of Tunisia. His collaborations with international institutions, such as Cairo University and Georg-August-Universität Göttingen, further highlight his commitment to advancing research in environmental sciences on a global scale. Through his extensive body of work, Azri continues to address critical environmental issues, contributing to the broader understanding of pollution and its effects on both ecosystems and human health.

Md. Badiuzzaman Khan is the Head of the Department of Environmental Science at Bangladesh Agricultural University (BAU). His research focuses on atmospheric aerosols, environmental chemistry, and pollution monitoring. Dr. Khan has contributed to over 16 international journal publications and has been recognized with several awards, including the Outstanding Reviewer Award from the *Journal of Environmental Chemical Engineering* in 2016. He is also a member of various professional organizations, such as the Science and Engineering Institute and the Indian Association for Air Pollution Control.

Environmental Monitoring with AI and IoT

Smart Systems for a Sustainable Future

Sneha Gautam, Chafai Azri, and
Md. Badiuzzaman Khan

CRC Press
Taylor & Francis Group
Boca Raton London New York

CRC Press is an imprint of the
Taylor & Francis Group, an **informa** business

Designed cover image: © Shutterstock

First edition published 2026
by CRC Press
4 Park Square, Milton Park, Abingdon, Oxon, OX14 4RN

and by CRC Press
2385 NW Executive Center Drive, Suite 320, Boca Raton FL 33431

© 2026 Sneha Gautam, Chafai Azri, and Md. Badiuzzaman Khan

CRC Press is an imprint of Informa UK Limited

British Library Cataloguing-in-Publication Data
A catalogue record for this book is available from the British Library

ISBN: 9781041196525 (hbk)
ISBN: 9781041196532 (pbk)
ISBN: 9781003712701 (ebk)

DOI: 10.1201/9781003712701

Typeset in Caslon
by codeMantra

Contents

Figures

Tables

Acknowledgments

We express our sincere gratitude to the contributors from around the world who brought diverse perspectives and invaluable insights to this work. Special thanks to our academic institutions (i.e., Karunya Institute of Technology and Sciences, India; University of Sfax, Tunisia and Bangladesh Agriculture University, Bangladesh) for supporting the development of this book volume. Finally, we acknowledge the dedicated editorial and publishing team whose guidance helped shape this book into a comprehensive and timely resource.

Introduction

Environmental monitoring is entering a new era. Traditional methods, while valuable, often lack the speed, scale, and precision required to address today's complex environmental challenges. The convergence of AI and IoT technologies provides an unprecedented opportunity to rethink how we observe, understand, and manage environmental systems.

This book explains the application of AI- and IoT-enabled systems to monitor the environment and climate in real time. It begins by introducing core concepts in advance tools (i.e., AI, Machine Learning, IoT sensor networks, and edge/cloud computing). The subsequent chapters explore domain-specific monitoring techniques, smart cities, data monitoring and analysis methods, digital twins, and policy implications. By exploring real-world global case studies and engaging in in-depth discussions of emerging tools, readers will acquire both the theoretical knowledge and practical skills necessary to design and implement intelligent environmental monitoring solutions.

PART I
FUNDAMENTAL CONCEPTS

1

INTRODUCTION TO ENVIRONMENTAL MONITORING

1.1 Definition, Scope, and Importance

Environmental monitoring refers to the systematic process of collecting, analyzing, and interpreting data to assess environmental conditions over time. This practice is essential for understanding the state of our natural surroundings, identifying potential environmental hazards, and ensuring the sustainability of ecosystems for future generations. By examining various environmental parameters such as air quality, water quality, biodiversity, climate patterns, and soil health, environmental monitoring enables comprehensive assessments of ecological integrity and human impact (Azri et al., 2009, Bahloul et al., 2015, Chabbi et al., 2017, 2021).

The scope of environmental monitoring is vast, encompassing critical areas like pollution control, conservation efforts, and public health protection (Azri, 2000; Gautam et al., 2016; Dunster et al., 2018; Hussain et al., 2023; Ahbil et al., 2024). It extends to monitoring greenhouse gas emissions, tracking biodiversity loss, evaluating the health of freshwater and marine ecosystems, and understanding the impacts of urbanization and industrial activities (Mkawar et al., 2007; Azri et al., 2010; Campbell et al., 2021; Baati et al., 2020; Ambade et al., 2023, Sellami et al., 2025). This multidimensional approach makes it a cornerstone of environmental science and management.

The importance of environmental monitoring is magnified in today's context of accelerating environmental challenges, such as climate change, deforestation, and the proliferation of pollutants due to industrial and urban expansion (Zaïbi et al. 2012; Dammak et al., 2016, 2020; Sellami et al., 2022, 2023). Accurate, real-time data derived from monitoring programs are indispensable for informed decision-making. Policymakers and regulatory authorities rely on

DOI: 10.1201/9781003712701-2

these data to craft effective policies and enforce environmental standards. For scientists, the data serve as a basis for research and innovation in sustainability practices. Furthermore, monitoring plays a crucial role in public health by identifying pollution sources, assessing exposure risks, and guiding interventions to reduce harmful impacts.

Additionally, environmental monitoring supports adaptive management strategies by providing evidence of trends and feedback on the effectiveness of mitigation measures. This ensures that conservation efforts remain dynamic and responsive to evolving environmental conditions. In essence, environmental monitoring acts as the eyes and ears of environmental stewardship, laying the foundation for a healthier planet and a sustainable future.

1.2 Traditional Monitoring Methods vs. Modern Technologies

Environmental monitoring has evolved significantly over time, transitioning from labor-intensive traditional methods to advanced, technology-driven approaches. Traditional environmental monitoring relied heavily on manual sampling, periodic measurements, and subsequent laboratory analysis. While these methods laid the foundation for understanding environmental conditions, they often came with limitations. The manual nature of sampling was resource-intensive, requiring significant time, effort, and specialized personnel. Data collection was typically periodic rather than continuous, leading to gaps in information and limited spatial and temporal coverage. As a result, traditional methods often struggled to provide the real-time insights necessary for rapid decision-making or immediate intervention during environmental crises. Modern technologies, however, have transformed this landscape by introducing automation, real-time capabilities, and advanced analytics. Innovations such as Artificial Intelligence (AI) and the Internet of Things (IoT) have redefined the scope and efficiency of environmental monitoring. AI, powered by Machine Learning algorithms, facilitates predictive analysis, allowing researchers to anticipate environmental trends and anomalies before they escalate into serious issues (Orru et al., 2017; Liu et al., 2022). It also excels at handling vast datasets, enabling the extraction of meaningful patterns and insights from complex environmental data.

IoT complements AI by providing the infrastructure for continuous, real-time data collection (Gope & Hwang, 2016; Tastan, 2018; Selvadass et al., 2022; Blessy et al., 2023).

Networks of interconnected sensors can monitor a wide range of environmental parameters from air and water quality to soil health and biodiversity, over expansive geographic areas. These sensors transmit data seamlessly, offering a level of temporal resolution and geographic coverage that was previously unattainable. For instance, IoT-enabled air quality monitors can detect minute changes in pollutant concentrations, while water sensors can provide live updates on contamination levels in rivers and lakes (Rasp et al., 2020; Liu et al., 2022; Packiavathy & Gautam, 2023). The integration of AI and IoT not only enhances the accuracy and granularity of environmental monitoring but also ensures that the data are actionable. Real-time analytics and dynamic dashboards enable policymakers, environmental managers, and researchers to make informed decisions quickly. Furthermore, these technologies reduce the reliance on manual labor and streamline the monitoring process, saving time and resources. In essence, while traditional methods have served as the cornerstone of environmental monitoring, modern technologies have propelled the field into a new era of efficiency and effectiveness. This shift empowers environmental management with tools capable of addressing the complex and dynamic challenges of the 21st century, fostering a more sustainable and resilient planet. Differences between traditional methods and modern technologies have been given in Table 1.1.

1.3 Overview of AI and IoT in Environmental Sciences

The integration of AI and the IoT has significantly advanced the field of environmental sciences, providing innovative tools to monitor, analyze, and manage the natural environment more effectively.

1.3.1 AI in Environmental Monitoring

AI focuses on leveraging computational power to process and interpret large volumes of environmental data, uncovering patterns, forecasting outcomes, and providing actionable insights. It excels in

Table 1.1   Traditional Monitoring Methods vs. Modern Technologies

ASPECT	TRADITIONAL MONITORING METHODS	MODERN TECHNOLOGIES	KEY EXAMPLES
Definition	Relies on manual or semi-automated techniques, such as field observations, laboratory analyses, and surveys, requiring direct human intervention (Xu et al., 2019).	Utilizes advanced tools like IoT sensors, Artificial Intelligence (AI), satellite imaging, and Big Data analytics for automated, real-time environmental monitoring (Rasp et al., 2020).	**Water Quality**: Lab-based water analysis vs. IoT sensors for real-time water quality monitoring. **Air Pollution**: Local devices vs. satellite imaging (Copernicus Climate Change Service, 2023).
Data collection	Involves periodic manual sampling and lab testing, which are time-consuming and labor-intensive (Ham et al., 2019).	Automated, real-time data collection using IoT devices, drones, and satellite remote sensing, enabling continuous monitoring (Liu et al., 2022).	**Biodiversity**: Manual species documentation vs. AI-powered drones tracking wildlife populations.
Accuracy	Prone to human errors and limited spatial-temporal resolution; results vary depending on data collection methods (Urban et al., 2020).	High precision with advanced algorithms, Machine Learning models, and tools that minimize errors and improve spatial-temporal coverage.	**Climate Modeling**: Historical trend analysis vs. AI-driven models combining real-time and historical data (Rasp et al., 2020).
Speed	Slow, with delayed results due to manual processing and lab testing, which may take days or weeks (Xu et al., 2019).	Provides near-instantaneous data processing and reporting through automated systems and cloud-based tools, enabling faster decision-making (Copernicus Climate Change Service, 2023).	**Disaster Prediction**: Slow manual compilation vs. real-time satellite imaging and predictive AI models.
Cost	Low initial costs but significant ongoing investments in manpower, equipment maintenance, and logistics.	High initial investment in technology but reduced operational costs over time due to scalability and automation (Alaba et al., 2017).	**Monitoring Costs**: Manual water sampling for local lakes vs. scalable IoT-enabled systems for nationwide monitoring.

(*Continued*)

Table 1.1 (*Continued*)   Traditional Monitoring Methods vs. Modern Technologies

ASPECT	TRADITIONAL MONITORING METHODS	MODERN TECHNOLOGIES	KEY EXAMPLES
Scalability	Limited scalability due to dependence on human resources, making large-scale or global projects costly and challenging.	Highly scalable, capable of handling large-scale applications like global climate monitoring, deforestation tracking, and urban air pollution control (NIST, 2021).	**Climate Monitoring:** Limited to specific sites using traditional methods vs. global-scale satellite and IoT systems.
Environmental impact	Minimal technological footprint but may involve invasive practices like water/soil sample removal and ecosystem disruption during surveys (Broy et al., 2020).	Non-invasive technologies like satellite imaging and drones reduce ecosystem disruption during data collection (ISO, 2021).	**Environmental Surveys:** Field sampling affecting ecosystems vs. drone-based remote sensing.
Accessibility	Constrained by geographic and weather-related limitations; requires skilled personnel and fieldwork (Jobin et al., 2019).	Accessible via internet-based platforms, cloud systems, and mobile apps, enabling remote operations even in hazardous locations (Liu et al., 2022).	**Remote Access:** Limited by weather/ geography in traditional methods vs. cloud-based data from remote devices globally.
Integration	Limited integration with other systems, requiring manual compilation and analysis of data from diverse sources (Ham et al., 2019).	Seamlessly integrates with digital tools such as GIS, AI, and Machine Learning platforms for automated analysis and visualization (Rasp et al., 2020).	**Data Analysis:** Standalone field datasets vs. integrated platforms combining GIS, AI, and Big Data analytics.
Applications	Used for specific studies like local water/air quality testing, biodiversity surveys, and climate data collection for specific locations (Urban et al., 2020).	Applied in diverse areas, including precision agriculture, disaster prediction, carbon tracking, and real-time climate modeling (Liu et al., 2022).	**Agriculture:** Local field soil testing vs. AI-powered tools for real-time monitoring in precision farming.

identifying complex relationships within environmental datasets, making it invaluable for trend analysis and predictive modeling. For instance, AI-powered Machine Learning algorithms can predict air quality fluctuations by analyzing historical pollution data, meteorological conditions, and urban activity patterns (Dua & Du, 2016;

Blumenstock, 2020; Dhyani & Kumar, 2021; Sarker, 2022). Similarly, neural networks can analyze vast, intricate datasets to detect correlations, such as the impacts of industrial emissions on biodiversity or climate patterns (Pan, 2016; Weiss et al., 2016; Shorten & Khoshgoftaar 2019; Wang et al., 2019; Alzubaidi et al., 2021). AI's ability to process unstructured data—such as satellite imagery or real-time video feeds—adds another layer of sophistication, enabling researchers to assess deforestation, glacier retreat, or urban sprawl with unprecedented precision. AI also plays a crucial role in resource optimization. Through advanced modeling, it can suggest strategies for efficient energy use, water resource allocation, or waste management. These capabilities make AI a cornerstone in environmental science, empowering stakeholders to make evidence-based decisions that balance ecological sustainability with human needs.

1.3.2 IoT in Environmental Monitoring

IoT, often described as the backbone of real-time environmental monitoring, connects physical devices to the digital realm. Sensors equipped with IoT technology continuously collect data on a wide range of environmental parameters, including air and water quality, temperature, humidity, soil health, and noise levels. These devices transmit the gathered data over networks to centralized platforms for processing and analysis. Unlike traditional monitoring methods, IoT devices operate seamlessly and remotely, eliminating the need for manual intervention and enabling near-continuous surveillance of environmental conditions. The real-time capabilities of IoT are especially critical in addressing environmental threats. For example, IoT-enabled water quality sensors can immediately detect contamination events, triggering alerts to prevent widespread harm. Similarly, air quality monitors can identify sudden pollution spikes, enabling authorities to take swift action to mitigate exposure risks. IoT devices also play a key role in monitoring remote or inaccessible locations, such as polar regions or dense forests, ensuring that no area remains unmonitored.

1.3.3 AI and IoT: A Synergistic Approach to Sustainability

When combined, AI and IoT create "smart systems" that redefine environmental management. IoT devices act as the data collection framework, while AI algorithms analyze and interpret these data to generate meaningful insights. Together, these technologies provide highly accurate, efficient, and scalable solutions (Alaba et al., 2017; Ham et al., 2019; Rasp et al., 2020; Liu et al., 2022). For example, a network of IoT sensors across an urban area can monitor real-time pollution levels, while AI models predict future air quality and recommend traffic or industrial adjustments to reduce emissions. The applications, future prospects, and challenges have been provided in Figure 1.1.

These smart systems enhance environmental management by offering dynamic and adaptive capabilities. They can monitor ecosystems continuously, predict environmental risks, and evaluate the success of interventions, ensuring resources are directed where they are most needed. As a result, AI and IoT empower policymakers, researchers, and communities with actionable information, enabling them to address environmental challenges more effectively and foster a more sustainable future. This synergy between AI and IoT marks a transformative step in environmental sciences, revolutionizing how we interact with and protect the natural world.

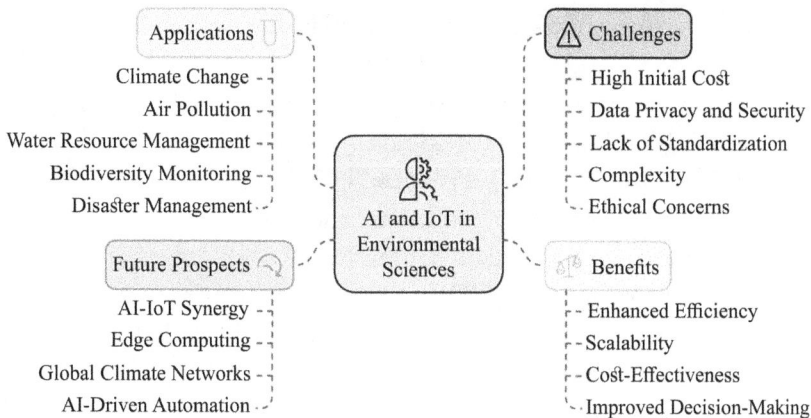

Figure 1.1   Overview of AI and IoT in environmental sciences (Alaba et al., 2017).

2

ENVIRONMENTAL PARAMETERS AND THEIR MEASUREMENT

2.1 Air Quality: Key Pollutants and Monitoring Techniques

Air quality is a fundamental environmental parameter with significant implications for human health, ecosystem stability, and climate dynamics. Poor air quality is associated with respiratory and cardiovascular diseases, biodiversity loss, and global warming, making its monitoring and management a critical priority.

2.2 Key Pollutants

The major pollutants that influence air quality include:

- **Particulate Matter (PM)**: Classified by size into PM1, PM2.5, and PM10, these tiny particles penetrate deeply into the respiratory system, causing severe health issues such as asthma, bronchitis, and cardiovascular disorders.
- **Nitrogen Oxides (NOx)**: Emitted primarily from vehicles and industrial processes, NOx contributes to the formation of ground-level ozone and acid rain.
- **Sulfur Dioxide (SO_2)**: Produced by burning fossil fuels, SO_2 aggravates respiratory conditions and contributes to the formation of fine particulate matter.
- **Carbon Monoxide (CO)**: A colorless, odorless gas from incomplete combustion of fuels that can be lethal at high concentrations.

 DOI: 10.1201/9781003712701-3

- **Ozone (O_3):** While beneficial in the stratosphere, ground-level ozone is a harmful pollutant formed through reactions between NOx and VOCs under sunlight.
- **Volatile Organic Compounds (VOCs):** Emitted from industrial activities, solvents, and vehicles, VOCs contribute to smog formation and pose health risks.

These pollutants originate from anthropogenic sources, such as vehicular emissions, industrial processes, and agricultural activities, as well as natural events like wildfires and volcanic eruptions.

2.3 Monitoring Techniques

Traditional air quality monitoring involved passive and active sampling methods. Passive samplers absorb pollutants over time, while active samplers draw air through filters for laboratory analysis. Although reliable, these methods offer limited spatial and temporal resolution and often require significant resources (Kaserzon et al., 2014; Jeong et al., 2018; Li et al., 2018; Jones et al., 2019). Modern air quality monitoring leverages advanced technologies, including real-time sensors and IoT-enabled devices (Mumtaz et al., 2021, Asha et al., 2023; Buelvas et al., 2023). These sensors continuously measure pollutant levels across multiple locations, providing comprehensive spatial and temporal data. AI algorithms enhance this process by analyzing sensor data to produce real-time air quality indices, predict pollution trends, and pinpoint hotspots of contamination. These "smart systems" enable authorities to take swift action, such as issuing public health advisories, enforcing industrial regulations, or implementing traffic control measures, to mitigate pollution exposure and reduce health risks.

2.4 Water Quality: Physical, Chemical, and Biological Indicators

Water quality is essential for human health, agriculture, industry, and the overall well-being of aquatic ecosystems. Monitoring water quality ensures the availability of safe drinking water, protects aquatic life, and maintains ecosystem balance.

2.5 Indicators of Water Quality

1. **Physical Indicators**

   These include measurable attributes like:
   - **Temperature**: Affects aquatic life and influences chemical reactions.
   - **Turbidity**: Measures water clarity; high turbidity indicates sedimentation or pollution.
   - **Color**: Can signal the presence of contaminants or organic materials.
   - **Flow Rate**: Impacts sediment transport and nutrient distribution.

   Monitoring physical indicators helps detect immediate environmental changes, such as algal blooms or sediment influx due to erosion.

2. **Chemical Indicators**

   These encompass substances dissolved in water, including:
   - **Dissolved Oxygen (DO)**: Essential for aquatic life; low DO levels indicate pollution.
   - **pH**: Reflects water's acidity or alkalinity, critical for aquatic organisms.
   - **Salinity**: Affects water usability for drinking and irrigation.
   - **Heavy Metals**: Toxic elements like lead and mercury can accumulate in the food chain.
   - **Nutrients**: Excessive nitrates and phosphates lead to eutrophication, degrading water quality.

3. **Biological Indicators**

   These assess the presence of microorganisms and aquatic organisms, such as:
   - **Bacteria**: Pathogens like *E. coli* signal contamination and pose health risks.
   - **Biodiversity**: The abundance and diversity of aquatic species provide insights into ecosystem health.

2.6 Monitoring Techniques

Traditional water quality monitoring relied on manual sample collection followed by laboratory analysis. While effective, this approach is often time-consuming, sporadic, and resource-intensive.

IoT-based monitoring systems have revolutionized water quality assessment by enabling continuous, real-time data collection. Sensors measure key water parameters like DO, pH, turbidity, and heavy metals, transmitting the data to centralized systems for analysis Figure 2.1.

AI applications further enhance water monitoring by analyzing trends, detecting contamination events, and predicting future water quality changes. For example, AI can identify pollution sources, such as agricultural runoff or industrial discharges, and model the potential impacts of these pollutants on water ecosystems.

By integrating IoT sensors and AI technologies, modern water quality monitoring ensures proactive management of water resources (Gopichand et al., 2024). These systems provide actionable insights for decision-makers, allowing for timely interventions to safeguard water supplies, protect ecosystems, and promote sustainability.

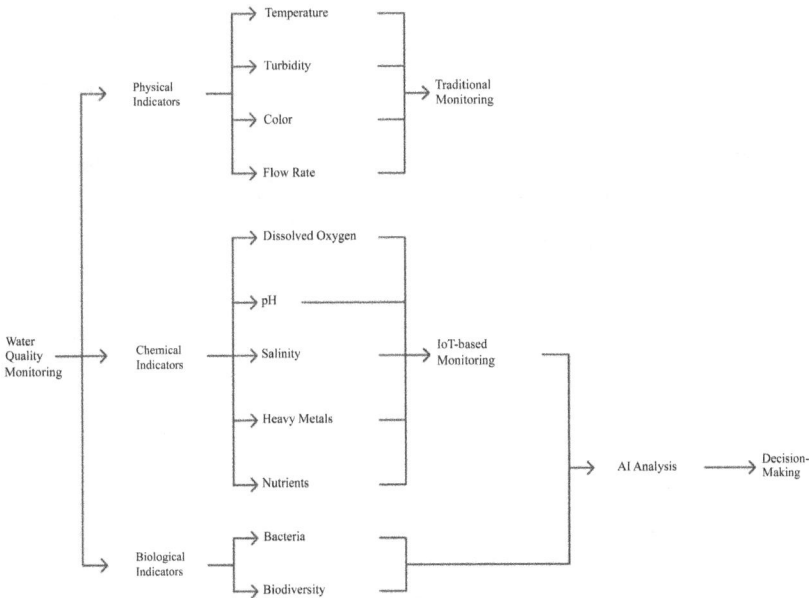

Figure 2.1   Applications of IoT-based monitoring for water quality parameters. (Created by the authors.)

2.7 Soil Health: Fertility, Contaminants, and Erosion Parameters

Soil health is essential for agricultural productivity, biodiversity, and ecosystem services. Key parameters include soil fertility, contamination levels, and erosion potential.

- **Soil Fertility**: Measured by the levels of key nutrients like nitrogen, phosphorus, potassium, and organic matter, soil fertility affects crop yields and the sustainability of agricultural practices.
- **Contaminants**: Heavy metals, pesticides, and industrial pollutants can degrade soil health. Monitoring soil for these contaminants helps in assessing the risk to human health and environmental stability.
- **Erosion Parameters**: Soil erosion is influenced by factors like rainfall intensity, land use, and soil composition. Monitoring soil erosion is essential to prevent land degradation and ensure sustainable farming practices.

IoT sensors and AI tools enable real-time soil monitoring, providing data on soil moisture, temperature, pH, and nutrient levels (Figure 2.2). Machine Learning models can predict soil degradation

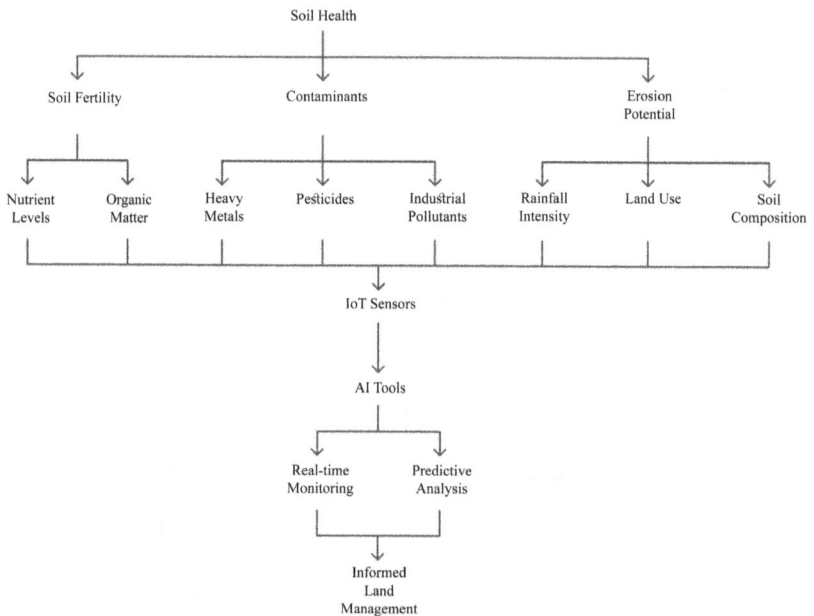

Figure 2.2   Applications of IoT-based monitoring for water quality parameters. (Created by the authors.)

and inform land management strategies, enhancing agricultural sustainability and mitigating erosion risks.

2.8 Climate Variables: Temperature, Humidity, and Precipitation

Climate variables such as temperature, humidity, and precipitation are key indicators of climate patterns and play a significant role in shaping ecosystems, agriculture, and human habitation.

- **Temperature**: It affects crop growth, water availability, and overall ecosystem health. Extreme temperature variations can signal changes in climate or environmental stress.
- **Humidity**: It influences evapotranspiration rates, agricultural productivity, and the likelihood of extreme weather events like heatwaves or storms.
- **Precipitation**: Regular monitoring of rainfall patterns is crucial for water resource management, agriculture, and disaster preparedness. Changes in precipitation levels can indicate shifts in regional climate patterns.

Traditional weather stations measure these climate variables, but IoT-enabled sensors now offer more localized and real-time monitoring. AI tools use these data to predict weather patterns, model climate changes, and inform decisions related to agriculture, water management, and disaster risk reduction. By integrating AI and IoT into the monitoring of air, water, soil, and climate parameters, environmental monitoring becomes more precise, timely, and actionable, helping to address environmental challenges in a sustainable manner.

3

FOUNDATIONS OF AI AND IoT FOR ENVIRONMENTAL APPLICATIONS

3.1 Basics of AI: Key Concepts and Algorithms

AI refers to the creation of intelligent systems capable of mimicking human cognitive functions such as reasoning, learning, problem-solving, and decision-making. In environmental science, AI is instrumental in processing large datasets, uncovering complex patterns, and making accurate predictions that manual analysis often cannot achieve.

3.2 Key Concepts in AI

1. **Machine Learning (ML)**

   ML is a core subset of AI that allows systems to learn and improve from experience without explicit programming. Using algorithms such as linear regression, decision trees, and random forests, ML models analyze historical and real-time data to identify trends, predict future conditions, and detect anomalies. For example, ML models can forecast air quality based on meteorological and pollutant data or detect unusual water contamination patterns to prompt early interventions.

2. **Neural Networks**

   Inspired by the human brain, neural networks consist of layers of interconnected nodes that process information. Advanced forms, such as deep learning networks, are highly effective in analyzing complex and high-dimensional datasets, including satellite images, environmental simulations, and sensor data. Neural networks are widely used for recognizing deforestation patterns, predicting climate anomalies, and classifying biodiversity in ecological studies.

DOI: 10.1201/9781003712701-4

3. **Natural Language Processing (NLP)**

   NLP enables AI systems to process, understand, and interpret human language. In environmental applications, NLP can analyze large volumes of unstructured data, such as scientific reports, policy documents, or public sentiment on social media, to extract insights about environmental awareness or regulatory compliance.

4. **Computer Vision**

   AI-driven computer vision analyzes visual data from images and videos, making it a valuable tool for monitoring biodiversity, detecting illegal logging, or assessing urban sprawl using satellite or drone imagery. This capability enhances environmental surveillance and supports conservation efforts.

3.3 AI Algorithms

AI employs a range of algorithms for classification, regression, clustering, and decision-making. Algorithms like support vector machines (SVMs), k-means clustering, and ensemble methods (e.g., gradient boosting) are extensively applied in environmental monitoring. These tools provide actionable insights by transforming raw data into meaningful information, supporting informed decision-making processes.

3.4 Internet of Things: Components, Architecture, and Applications

IoT is a network of interconnected physical devices embedded with sensors, software, and communication technologies that enable them to collect, share, and act on environmental data. IoT systems form the backbone of modern environmental monitoring, providing comprehensive and real-time insights.

3.5 Key Components of IoT

1. **Sensors**

   Sensors are the primary data collection tools in IoT, monitoring parameters such as temperature, humidity, air and water quality, and soil moisture. These sensors deliver accurate and continuous data, essential for understanding dynamic environmental conditions.

2. **Connectivity**

   IoT devices rely on communication networks to transmit collected data. Technologies like Wi-Fi, Bluetooth, Zigbee, and cellular networks are used for localized monitoring, while Low Power Wide Area Networks (LPWANs) such as LoRaWAN are ideal for large-scale deployments in remote areas.

3. **Data Processing and Storage**

   Collected data must be processed and stored, either in cloud-based platforms for centralized access or via edge computing for localized analysis. This dual approach ensures real-time responsiveness and scalability.

4. **Actuators**

   Actuators are components that take physical actions based on sensor data. In environmental applications, they are used for automation, such as adjusting irrigation levels in response to soil moisture data or managing ventilation systems to optimize indoor air quality.

3.6 Applications of IoT

IoT applications in environmental sciences include real-time air and water quality monitoring, smart waste management, and energy optimization. For instance, IoT systems can monitor pollution across large urban areas, provide live updates, and support the development of targeted mitigation strategies.

3.7 Synergies between AI and IoT for Environmental Solutions

The combination of AI and IoT represents a transformative approach to environmental monitoring and management, leveraging real-time data collection, advanced analytics, and automated decision-making. Key benefits of the integration of AI and IoT have been given in Table 3.1.

Table 3.1   Synergies between AI and IoT for Environmental Solutions

CATEGORY	DESCRIPTION	EXAMPLES
Definition	The integration of AI and IoT combines real-time data collection (IoT) with data analysis and decision-making capabilities (AI), enabling smarter and faster environmental solutions (Liu et al., 2022).	AI analyzing IoT-collected air quality data to predict pollution trends and suggest mitigation measures.
Data processing	IoT collects environmental data from sensors (e.g., air quality, water quality, temperature) in real time, while AI processes these data using advanced algorithms for trend analysis, anomaly detection, and predictions (Rasp et al., 2020).	IoT-enabled smart water meters feeding data to AI models to predict water consumption patterns and optimize usage in urban areas.
Predictive analytics	AI-powered predictive analytics combined with IoT data streams help forecast environmental events such as extreme weather, droughts, or flooding (Ham et al., 2019).	Flood forecasting using IoT-based sensors in rivers and AI to analyze rainfall patterns and water flow.
Automation	The synergy enables automated environmental monitoring systems, reducing the need for human intervention in repetitive tasks like pollution monitoring and resource allocation (Urban et al., 2020).	Automated smart irrigation systems that adjust water supply based on soil moisture data from IoT sensors and AI weather predictions.
Real-time decision-making	AI processes IoT data instantaneously, enabling timely responses to environmental hazards such as industrial spills, wildfires, or rising air pollution levels (Alaba et al., 2017).	AI-driven wildfire monitoring systems using IoT-enabled heat sensors for immediate alerts and action.
Scalability	The combined systems are highly scalable, capable of expanding to large-scale applications like global climate monitoring and city-wide smart systems (Copernicus Climate Change Service, 2023).	Integration of IoT sensors in global deforestation monitoring networks with AI-powered image recognition for real-time reporting.
Applications	• Climate change monitoring • Water resource management • Pollution control • Biodiversity conservation • Disaster response (ISO, 2021).	IoT-enabled biodiversity tracking using AI to analyze animal migration patterns. Smart city air quality systems with real-time alerts.

(Continued)

Table 3.1 (*Continued*)   Synergies between AI and IoT for Environmental Solutions

CATEGORY	DESCRIPTION	EXAMPLES
Benefits	• **Efficiency**: Automation reduces human error. • **Real-Time Insights**: Combines IoT data with AI analysis for faster decisions. • **Cost Savings**: Long-term operational cost reductions. • **Predictive Maintenance**: Avoids equipment failures. • **Sustainability**: Promotes optimized resource use.	Smart irrigation, AI-based disaster prediction, and smart city energy systems.
Challenges	• **High Costs**: Expensive to develop and deploy. • **Data Privacy Issues**: Risk of sensitive environmental data being compromised. • **Complexity**: Requires expertise for integration. • **Lack of Standards**: Inconsistent device compatibility. • **Ethical Concerns**: Potential misuse of technology.	IoT systems hacked to alter data; inconsistent IoT sensors in biodiversity monitoring.

3.8 Key Benefits of AI-IoT Integration

1. Real-Time Data Processing and Analysis

IoT sensors continuously generate data streams that AI algorithms analyze to identify patterns, detect anomalies, and provide actionable insights. For example, in air quality monitoring, IoT devices measure pollutant levels, while AI predicts pollution trends and suggests mitigation strategies.

2. Predictive Capabilities

AI, when integrated with IoT, enables forecasting of environmental events such as pollution spikes, floods, or droughts. This predictive power allows for proactive measures, such as issuing early warnings or optimizing resource allocation to minimize damage.

3. Automated Decision-Making

AI-powered systems can automate responses based on IoT sensor inputs. For instance, in a smart irrigation system, AI can analyze soil moisture data and control water flow, ensuring efficient use and minimizing waste.

4. **Optimization and Efficiency**

   AI and IoT together optimize resource utilization by identifying inefficiencies and suggesting improvements. In energy management, IoT sensors monitor consumption, while AI algorithms recommend distribution adjustments to reduce wastage.

5. **Scalability**

   IoT's wide coverage, combined with AI's ability to process data at scale, enables solutions that address both localized and global environmental challenges. These technologies are vital for managing complex issues like climate change, urban air quality, and water scarcity.

By synergizing AI and IoT, environmental monitoring systems become smarter, more efficient, and highly responsive, empowering decision-makers with real-time insights and proactive tools to build a sustainable future.

4
ROLE OF AI IN ENVIRONMENTAL MONITORING AND SUPPORTING TECHNOLOGIES

Artificial Intelligence (AI) is becoming an essential tool in modern environmental surveillance, providing efficient, cost-effective, and accurate ways to study natural systems and guide their stewardship. This proposal emphasizes leveraging AI technologies to convert unprocessed data into meaningful insights-supporting proactive alert mechanisms, data-driven policymaking, and the responsible use of environmental resources.

1. **Image Analysis and Remote Sensing**

   Convolutional neural networks are widely used in processing satellite and drone images for various applications such as land use classification, deforestation tracking, glacier monitoring, and heat island detection. These models can identify subtle changes over time that traditional methods often miss.

2. **Time Series Forecasting**

   Recurrent neural networks and long short-term memory networks are effective for predicting environmental trends such as rainfall, drought, floods, and pollution. These forecasts are essential for risk mitigation and emergency planning.

3. **Real-Time Monitoring with IoT and Edge AI**

   By combining AI with smart sensor networks (IoT), real-time environmental monitoring is possible even in remote or harsh regions. Edge AI processes data locally, reducing latency and dependence on cloud services. It is applied in monitoring air and water quality, noise pollution, and soil conditions.

DOI: 10.1201/9781003712701-5

4. **Environmental Anomaly Detection**

Unsupervised Machine Learning models like clustering algorithms and autoencoders are useful for detecting anomalies or irregular patterns in large datasets—helpful for identifying illegal mining, poaching, invasive species, or unanticipated pollution.

5. **Climate Modeling and Simulation**

AI enhances the resolution and accuracy of climate models by capturing complex data relationships. AI-based downscaling techniques bridge the gap between global projections and localized predictions.

6. **Biodiversity and Species Tracking**

AI models trained on acoustic and visual data from bioacoustic sensors or camera traps can accurately identify species presence, population trends, and behaviors. These tools are vital for conservation, particularly in biodiversity-rich regions.

7. **Decision Support and Policy Tools**

AI-powered systems integrate predictive modeling with optimization algorithms to support strategic planning—such as optimizing irrigation, preventing wildfires, or designing emission reduction pathways.

8. **Citizen Science and Crowdsourced Data**

AI helps process and validate large datasets submitted through citizen science apps. Natural language processing and image recognition models improve the consistency and scientific usability of the collected data.

9. **Big Data and Cloud Integration**

Environmental data are often vast, unstructured, and varied. AI depends on robust cloud-based pipelines for storage, processing, and visualization. These systems allow for interactive dashboards and real-time reporting, promoting collaboration and transparency.

AI-based technologies do more than just track the environment—they foster a proactive, sustainable approach aligned with the United Nations Sustainable Development Goals, particularly Goals 13 (Climate Action), 14 (Life Below Water), and 15 (Life on Land).

PART II
CORE TECHNOLOGIES AND METHODOLOGIES

5

SENSOR TECHNOLOGIES IN ENVIRONMENTAL MONITORING

5.1 Types of Sensors for Air, Water, and Soil

Sensors play an indispensable role in environmental monitoring by providing accurate, real-time data on critical environmental parameters (Chen et al., 2016; Alam, 2017; Gupta et al., 2017; Ramya et al., 2020). These devices are tailored to measure specific physical, chemical, or biological properties, enabling effective management of air, water, and soil resources.

5.2 Air Quality Sensors

Air quality sensors detect and quantify pollutants that significantly impact human health and ecosystems.

- **Key Parameters Monitored**: Particulate matter (PM1, PM2.5, PM10), nitrogen oxides (NOx), sulfur dioxide (SO_2), carbon monoxide (CO), carbon dioxide (CO_2), ozone (O_3), and volatile organic compounds (VOCs).
- **Technologies Used**:
    - **Electrochemical Sensors**: Ideal for detecting gases like NOx, SO_2, and CO by measuring changes in electrical conductivity due to gas interactions.
    - **Metal-Oxide Semiconductor (MOS) Sensors**: Used for detecting VOCs and other gases through surface reactions that alter electrical resistance.
    - **Optical Sensors**: Measure particulate matter using light scattering techniques to determine particle sizes and concentrations.

DOI: 10.1201/9781003712701-7

- **Applications**: These sensors are integral to air quality monitoring networks, smart city initiatives, and research studies focused on pollution impacts on health and climate.

5.3 Water Quality Sensors

Water quality sensors evaluate the physical, chemical, and biological characteristics of water to ensure its safety and ecological integrity.

- **Key Parameters Monitored**: pH, dissolved oxygen (DO), turbidity, conductivity, temperature, and concentrations of contaminants like heavy metals, nitrates, and phosphates.
- **Technologies Used**:
  - **Optical Sensors**: Measure turbidity and detect contaminants through light absorption and scattering.
  - **Ion-Selective Electrodes (ISE)**: Specialized sensors for detecting specific ions such as nitrates or phosphates in water.
  - **Electrochemical Sensors**: Used for monitoring pH, DO, and conductivity by measuring electrical properties of the water.
- **Applications**: Water quality sensors are essential in managing drinking water safety, monitoring aquatic ecosystems, and detecting contamination events in industrial and agricultural runoff.

5.4 Soil Quality Sensors

Soil sensors provide data on the physical and chemical properties of soil, which are critical for agriculture, forestry, and land management.

- **Key Parameters Monitored**: Soil moisture, temperature, pH, salinity, and nutrient levels (e.g., nitrogen, phosphorus, potassium).
- **Technologies Used**:
  - **Capacitive and Resistive Moisture Sensors**: Measure soil moisture by detecting changes in dielectric constant or electrical resistance.

- **Electrochemical Sensors**: Used for pH and nutrient analysis through ion-sensitive membranes.
- **Optical Sensors**: Measure soil composition by analyzing light reflectance or absorption in soil samples.
- **Applications**: These sensors support precision agriculture by optimizing irrigation and fertilization practices, improving crop yields, and maintaining soil health.

5.5 Significance and Applications

Each type of sensor provides essential data that support informed decision-making in environmental management. From mitigating air pollution and ensuring water safety to enhancing agricultural productivity and conserving ecosystems, these sensors are critical to achieving sustainability goals (Chen et al., 2017; Gupta et al., 2017). Modern advancements in sensor technologies, including their integration with IoT and AI, have enhanced their accuracy, efficiency, and ability to provide actionable insights in real time.

5.6 Principles of Sensor Operation

The functionality of environmental sensors is grounded in the principles of physics, chemistry, and biology, enabling them to detect and measure various environmental parameters accurately (Copernicus Climate Change Service, 2023; Gopichand et al., 2024). Different sensing mechanisms are employed depending on the specific conditions or pollutants to be monitored (Figure 5.1).

5.6.1 Optical Sensing

Optical sensing involves the use of light to detect and analyze environmental parameters.

- **Working Principle**: Light interacts with particles or substances, either being scattered, absorbed, or reflected, which is then measured to determine the parameter of interest.
- **Applications**:
  - **Air Quality Monitoring**: Optical sensors measure particulate matter (PM1, PM2.5, PM10) concentrations by

Figure 5.1 Principles of sensor operation (Copernicus Climate Change Service, 2023).

analyzing the scattering of light when particles are present in the air.

- **Water Quality Monitoring**: Turbidity, an indicator of water clarity, is measured by evaluating the absorption and scattering of light caused by suspended solids.
- **Advantages**: Non-invasive, precise, and capable of providing real-time data.

5.6.2 Electrochemical Sensing

Electrochemical sensors detect environmental pollutants through chemical reactions.

- **Working Principle**: These sensors consist of electrodes that interact with specific gases or dissolved substances, triggering a chemical reaction that produces an electrical signal proportional to the pollutant concentration.
- **Applications**:
  - **Air Monitoring**: Detection of gases like CO_2, NOx, SO_2, and CO.
  - **Water Quality Monitoring**: Measurement of DO to assess aquatic ecosystem health.
- **Advantages**: High sensitivity to specific substances and effective in detecting low concentrations.

5.6.3 Capacitive and Resistive Sensing

These methods are commonly used to measure moisture content in soil and water.

- **Working Principle**:
  - **Capacitive Sensors**: Detect changes in the dielectric constant caused by the presence of water.
  - **Resistive Sensors**: Measure the resistance of the medium, which decreases with increasing moisture.
- **Applications**:
  - **Soil Quality Monitoring**: Assess soil moisture levels to optimize irrigation in agriculture.
  - **Water Management**: Monitor water retention in soil for hydrological studies.
- **Advantages**: Simple design, cost-effective, and suitable for continuous monitoring.

5.6.4 Ion-Selective Electrodes (ISE)

ISE sensors are specialized for detecting specific ions in water.

- **Working Principle**: A selective membrane interacts with target ions (e.g., nitrates, phosphates, potassium), generating a voltage proportional to their concentration.
- **Applications**:
  - **Water Quality Monitoring**: Detect nutrient levels in water, helping to prevent eutrophication.
  - **Agriculture**: Assess the availability of key nutrients in soil for crop health.
- **Advantages**: High specificity and rapid detection of target ions.

5.6.5 Biological Sensing

Biological sensors use living organisms or biological reactions to assess environmental conditions.

- **Working Principle**: These sensors detect the presence of specific organisms or measure biological responses, such as microbial activity, that indicate environmental changes.
- **Applications**:
  - **Water Monitoring**: Detect pathogens like *E. coli* to ensure water safety.
  - **Ecosystem Health**: Evaluate biodiversity and microbial activity in soil or water to monitor ecosystem integrity.
- **Advantages**: Unique capability to assess biological and ecological health.

5.7 Choosing the Right Sensing Principle

Each sensing principle is tailored to the specific environmental parameter being measured, balancing factors such as accuracy, sensitivity, cost, and operational complexity. This precision ensures reliable data collection, forming the foundation for effective environmental monitoring and informed decision-making. Modern advancements in sensor technology, coupled with AI and IoT, further enhance these capabilities, enabling dynamic, real-time environmental assessments.

5.8 Calibration, Accuracy, and Maintenance of Sensors

Environmental monitoring systems rely heavily on the proper functioning of sensors to provide accurate and reliable data. To ensure the precision and consistency of the readings over time, sensors must be calibrated, maintained, and regularly checked for accuracy.

5.8.1 Calibration

Calibration is the process of adjusting a sensor's output to ensure its readings match a known reference value or standard. Without proper calibration, sensors may provide inaccurate or unreliable data, which could have serious implications for environmental monitoring efforts.

- **Purpose of Calibration**: Calibration ensures that the sensor's measurements are consistent and accurate over time, allowing it to provide trustworthy data for decision-making in areas like air and water quality monitoring.
- **Process of Calibration**: Calibration is typically performed by exposing the sensor to a known standard or reference material. For example:
  - **Air Quality Sensors**: A sensor designed to measure pollutants like O_3 or nitrogen dioxide (NO_2) would be calibrated using known concentrations of these pollutants to verify that the sensor is accurately detecting and reporting their levels.
  - **Water Quality Sensors**: In water quality monitoring, calibration may involve using reference solutions with known pH levels or concentrations of specific ions, such as nitrates or phosphates, to ensure the sensor accurately measures water parameters.
- **Frequency of Calibration**: Sensors may need recalibration periodically or after specific events like sensor maintenance, extreme environmental changes, or after the sensor has been exposed to conditions outside its optimal operating range.

5.8.2 Accuracy

The accuracy of a sensor refers to how close its measurements are to the true value of the environmental parameter it is designed to monitor. Accurate measurements are essential for reliable environmental data that can inform important decisions about public health and ecosystem preservation.

- **Factors Affecting Accuracy**:
  - **Sensor Drift**: Over time, the performance of sensors can degrade due to environmental conditions such as temperature fluctuations, humidity, and chemical exposure. This is known as "sensor drift."
  - **Environmental Conditions**: Variations in environmental conditions, such as extreme temperatures or high levels of pollutants, can also affect sensor accuracy.
  - **Quality of the Sensor**: High-quality sensors are typically more accurate and stable, but even they require proper maintenance and calibration.
- **Maintaining Accuracy**: To maintain accuracy, it is essential to regularly calibrate sensors and compare their measurements with those of reference instruments. For example, if a sensor is measuring air quality, its readings should be cross-checked with data from government-regulated air quality monitoring stations to verify that it remains accurate.

5.8.3 Maintenance

Routine maintenance is a crucial part of sensor management to ensure sensors continue functioning optimally over time. Proper maintenance can help extend the lifespan of sensors, reduce errors, and ensure high-quality data collection.

- **Maintenance Tasks**:
  - **Cleaning**: Environmental sensors, especially those used in air or water monitoring, may accumulate dust, debris, or pollutants that can interfere with their performance. Regular cleaning ensures that the sensor's surface is clear of obstructions.

- **Replacing Components**: Some sensors may require the replacement of components such as electrodes, filters, or batteries, which wear out over time due to prolonged exposure to environmental conditions.
- **Physical Inspection**: Sensors should be regularly checked for physical damage, such as cracks or corrosion, that could compromise their operation.
- **Recalibration**: Sensors should be recalibrated periodically, especially after maintenance tasks or if any sensor component has been replaced.
- **Frequency of Maintenance**: The frequency of maintenance depends on factors such as the sensor type, environmental exposure (e.g., temperature extremes, humidity), and the importance of the monitoring system. For example, sensors used in critical air or water quality monitoring systems may require more frequent maintenance than those used in less demanding applications.

5.8.4 Sensor Lifetime

Over time, all sensors will experience degradation in performance due to prolonged exposure to harsh environmental conditions. Factors such as temperature fluctuations, high humidity, and exposure to contaminants can shorten a sensor's effective operational life.

- **Monitoring Performance**: It is essential to monitor the sensor's performance over time, including comparing its output to known reference values, to detect any signs of degradation.
- **End of Life**: When a sensor no longer provides reliable or accurate readings, it may be time to replace it. This could be due to the natural wear and tear of components or the sensor reaching the end of its functional lifespan.
- **Replacement**: Regularly replacing sensors ensures the continuity of accurate environmental monitoring, which is critical for assessing and addressing environmental issues in real-time.

Proper calibration, regular maintenance, and monitoring of sensor accuracy are essential to ensuring that environmental sensors continue to provide reliable, actionable data. The quality of the data obtained from these sensors directly impacts the effectiveness of environmental monitoring and management strategies aimed at preserving natural resources, public health, and ecosystems. By maintaining sensors in peak condition, environmental monitoring systems can be optimized for long-term, reliable use, contributing to sustainable development and more informed decision-making.

6

IoT Infrastructure and Communication Protocols

6.1 Network Architecture for IoT Systems

The network architecture of Internet of Things (IoT) systems forms the foundation for connecting, collecting, and processing environmental data from a range of devices. It is composed of several interconnected layers and components that work in harmony to enable real-time monitoring and decision-making. Each layer plays a distinct role in ensuring the efficiency, scalability, and security of the IoT ecosystem (Abbas & Yoon, 2015). The network architecture of IoT systems and their applications with networking parts have been given in Figure 6.1.

6.1.1 Sensors/Devices: The Data Generation Layer

Sensors and devices represent the initial touchpoint in the IoT network, where environmental data are captured.

- **Role of Sensors**: Sensors are deployed in the field to monitor specific environmental parameters such as air quality, water quality, soil health, temperature, and humidity.
- **Examples in Environmental Monitoring**:
  - **Air Quality Sensors**: Detect particulate matter, gases (NO_2, SO_2, CO_2), and volatile organic compounds.
  - **Water Quality Sensors**: Measure pH, turbidity, conductivity, and dissolved oxygen levels in water bodies.
  - **Soil Moisture Sensors**: Assess soil water content to optimize agricultural irrigation practices.
- **Importance**: This layer forms the backbone of the IoT network, providing real-time data streams that reflect the current state of the environment.

DOI: 10.1201/9781003712701-8   **37**

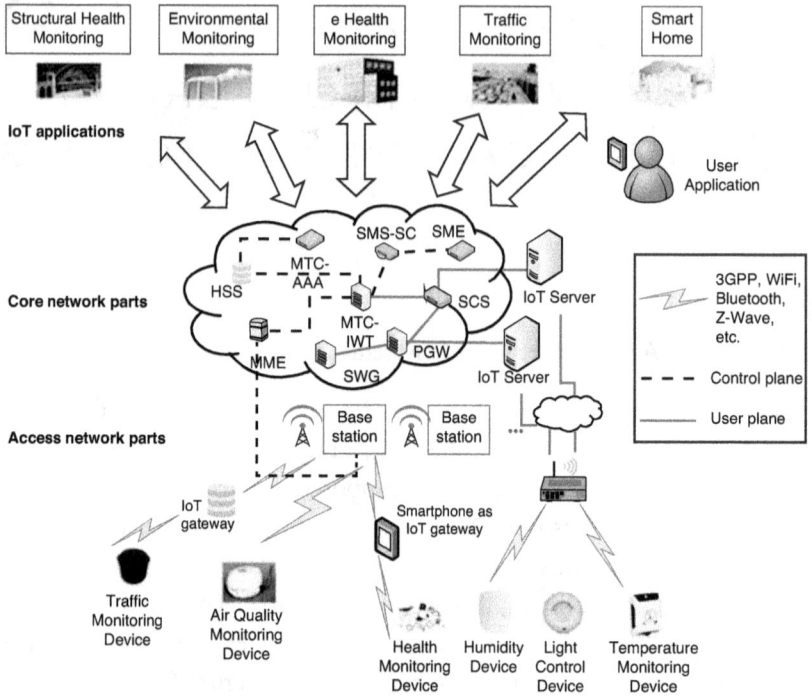

Figure 6.1 The network architecture of IoT systems and their applications (Abbas & Yoon, 2015).

6.1.2 Gateways: The Data Aggregation and Transmission Layer

Gateways serve as the bridge between sensors/devices and higher layers of the IoT network, ensuring smooth data flow and preliminary processing.

- **Role of Gateways**
  - Aggregate raw data from multiple sensors.
  - Perform data preprocessing tasks such as filtering, validation, and normalization to remove noise or errors.
  - Secure data through encryption protocols to protect it during transmission.
  - **Compression**: Reduce the size of data packets to optimize transmission efficiency.
- **Communication Technologies**: Gateways facilitate communication between devices using various protocols such as Wi-Fi, Zigbee, LoRaWAN, or cellular networks (e.g., 4G/5G).

- **Importance**: By consolidating and preparing data, gateways minimize the load on downstream systems, ensuring seamless data flow.

6.1.3 Cloud and Edge Computing: The Data Processing Layer

IoT systems leverage cloud and edge computing to process and analyze data efficiently, depending on specific use cases.

- **Cloud Computing**:
  - Provides centralized platforms for large-scale data processing, long-term storage, and advanced analytics.
  - Facilitates integration with Artificial Intelligence (AI) and Machine Learning algorithms to uncover insights and predictions from historical and real-time data.
  - Suitable for applications requiring extensive storage and computational power, such as climate modeling or historical trend analysis.
- **Edge Computing**:
  - Processes data closer to its source—near the sensors or gateways—reducing latency and enhancing response times.
  - Ideal for real-time applications like air quality alarms or automated irrigation systems, where immediate action is required.
- **Hybrid Approach**: Many IoT systems adopt a hybrid model, using edge computing for real-time decision-making and cloud computing for in-depth analysis and archiving.
- **Importance**: These systems enable the rapid processing of large data volumes, optimizing performance based on use-case demands.

6.1.4 Data Storage and Analysis: The Decision-Making Layer

Data transmitted to the cloud or edge servers are stored and analyzed to provide actionable insights.

- **Data Storage**: IoT systems employ scalable storage solutions such as relational databases, NoSQL databases, or data lakes to manage diverse and high-volume datasets.
- **Data Analysis**: Advanced analytics techniques, often powered by AI and Machine Learning, are used to extract meaningful insights. Examples include:
  - **Predictive Analytics**: Forecasting pollution levels or water shortages based on historical trends.
  - **Real-Time Alerts**: Triggering alarms when sensor readings exceed safe thresholds, such as during a chemical spill or air quality deterioration.
  - **Automated Decisions**: For instance, activating irrigation systems when soil moisture levels drop below optimal levels.
- **Visualization Tools**: Dashboards and data visualization platforms help present complex data in an accessible format, aiding stakeholders in decision-making.

6.1.5 Layered Architecture in IoT Systems

IoT systems typically operate in a **layered architecture**, where each layer performs a specific function:

1. **Perception Layer (Sensors)**: Captures environmental data.
2. **Network Layer (Gateways)**: Facilitates secure and efficient data transmission.
3. **Processing Layer (Cloud/Edge)**: Analyzes and stores data.
4. **Application Layer (Insights)**: Provides user interfaces and actionable insights for decision-making.

The network architecture of IoT systems integrates a wide array of technologies and components to ensure seamless data acquisition, transmission, processing, and analysis. By structuring the system in well-defined layers, IoT solutions can achieve scalability, reliability, and efficiency, enabling the proactive management of environmental challenges. This architecture not only facilitates real-time monitoring but also empowers stakeholders with insights to make informed decisions for sustainability and resource conservation.

6.2 Data Transmission: Wi-Fi, LPWAN, Zigbee, and 5G

Efficient data transmission is a cornerstone of IoT systems, ensuring that real-time environmental data collected by sensors are communicated seamlessly for processing and decision-making. Different communication protocols cater to diverse requirements such as range, power efficiency, data transfer speed, and network coverage. Here is an in-depth look at the key transmission technologies:

6.2.1 Wi-Fi: High-Speed Connectivity for Localized Networks

- **Overview**: Wi-Fi is one of the most commonly used communication protocols in IoT systems where a stable internet connection is available.
- **Key Features**:
  - High data transmission rates, suitable for applications requiring large bandwidth.
  - Operates within a limited range (typically up to 100 meters in indoor settings).
  - Standardized on IEEE 802.11 protocols, ensuring compatibility with a wide range of devices.
- **Use Cases**:
  - Environmental monitoring in urban areas with strong network infrastructure.
  - Air quality sensors transmitting high-resolution data in real time.
- **Challenges**:
  - High power consumption makes it less suitable for battery-powered devices.
  - Limited scalability for expansive rural or remote deployments.

6.2.2 Low-Power Wide-Area Networks (LPWAN):
Long-Range, Low-Power Solutions

- **Overview**: LPWAN technologies, including LoRaWAN and Sigfox, are designed to provide efficient communication over long distances while consuming minimal power.
- **Key Features**:
  - Exceptional range, often covering several kilometers.
  - Low data rate, optimized for transmitting small packets such as sensor readings.
  - Operates on unlicensed frequency bands, reducing operational costs.
- **Use Cases**:
  - Remote environmental monitoring in forests, agricultural fields, and water bodies.
  - Air and water quality monitoring in areas with limited infrastructure.
- **Advantages**:
  - Extended battery life, enabling sensors to function for years without frequent maintenance.
  - Scalability for wide-area deployments.
- **Challenges**:
  - Limited bandwidth makes it unsuitable for applications requiring high-speed data transfer.

6.2.3 Zigbee: Energy-Efficient Communication for Local Sensor Networks

- **Overview**: Zigbee is a short-range protocol tailored for low-power, low-data-rate applications in confined environments.
- **Key Features**:
  - Based on the IEEE 802.15.4 standard, it is optimized for short-distance communication (typically under 100 meters).
  - Forms mesh networks, allowing devices to relay data, which improves coverage and reliability.
- **Use Cases**:
  - Localized environmental monitoring systems, such as indoor air quality or soil moisture sensors.

- Smart agriculture setups where sensors are deployed within close proximity.
- **Advantages**:
  - Extremely low power consumption, suitable for battery-operated sensors.
  - Cost-effective for small-scale deployments.
- **Challenges**:
  - Limited range and data rate compared with Wi-Fi or 5G.
  - Best suited for networks with relatively simple data requirements.

6.2.4 5G: The Future of High-Speed IoT Connectivity

- **Overview**: The rollout of 5G networks is transforming IoT communications by delivering unprecedented speed, capacity, and reliability. Figure 6.2 illustrated ambient IoT devices introduced by 5G advanced that harvest energy from radio frequency waves.
- **Key Features**:
  - Ultra-low latency, ensuring near-instantaneous data transmission.

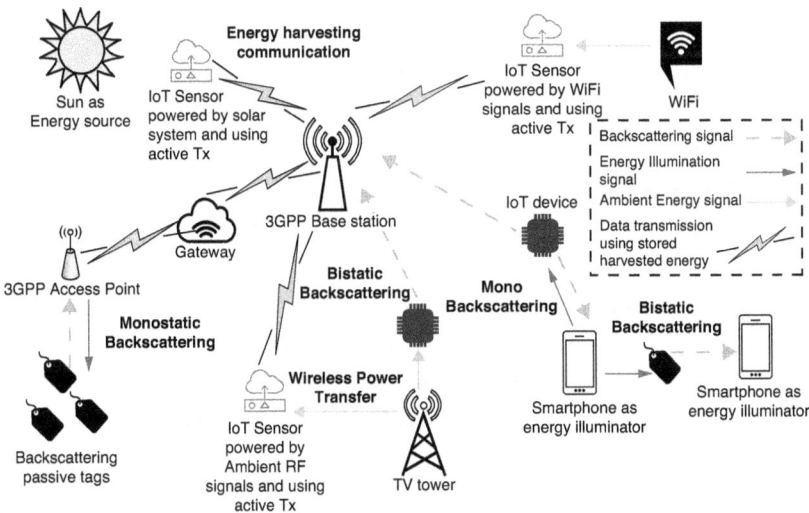

Figure 6.2 Diagram illustrating various methods of energy harvesting for ambient IoT devices (Majid Butt et al., 2023).

- High data rates, enabling transmission of large datasets in real time.
- Supports massive device connectivity, accommodating millions of sensors within a single network.
- **Use Cases:**
  - Urban air quality monitoring with high-resolution data streams.
  - Disaster response systems requiring real-time communication for decision-making.
  - Smart city applications integrating environmental, traffic, and infrastructure monitoring.
- **Advantages:**
  - Seamless support for complex, data-intensive applications.
  - Enables cutting-edge applications such as predictive analytics and AI-driven insights.
- **Challenges:**
  - Infrastructure limitations in rural or remote areas.
  - Higher operational costs compared to other protocols.

6.3 Choosing the Right Protocol: Balancing Needs and Constraints

The selection of a communication protocol depends on multiple factors:

- **Distance and Coverage:** LPWAN is ideal for long-range rural monitoring, while Wi-Fi and Zigbee excel in localized networks.
- **Power Consumption:** Zigbee and LPWAN are better suited for battery-powered devices due to their low energy requirements.
- **Data Rate:** Applications requiring high-speed data, such as urban air quality monitoring, benefit from Wi-Fi or 5G.
- **Environmental Conditions:** Rugged and remote environments often necessitate LPWAN due to its reliability and range.

By leveraging the strengths of these communication protocols, IoT systems can be tailored to meet the unique demands of environmental monitoring, ensuring robust, efficient, and scalable solutions.

6.4 Challenges in IoT Deployment for Environmental Applications

The deployment of IoT technology in environmental monitoring systems has vast potential to revolutionize the way we monitor and manage environmental parameters such as air quality, water quality, and soil health. However, several challenges must be overcome to ensure the success and efficiency of IoT systems in these applications:

6.4.1 Network Connectivity: Overcoming Remote Area Barriers

- **Challenge**: Reliable network connectivity is essential for the smooth operation of IoT systems. In many rural or remote areas, establishing high-quality communication infrastructure can be difficult. Poor cellular or internet coverage limits the ability to transmit real-time data from IoT sensors to cloud storage or processing units.
- **Solution**: While LPWANs such as LoRaWAN and Sigfox can help bridge this gap by offering long-range communication with minimal infrastructure, network coverage remains a significant challenge. Solutions like satellite communication or localized mesh networks may provide alternatives, but infrastructure limitations still persist.
- **Impact**: Poor connectivity may hinder the effectiveness of IoT systems in remote environmental monitoring and delay the acquisition of critical environmental data.

6.4.2 Power Consumption: Meeting the Demands of Remote Monitoring

- **Challenge**: Many environmental monitoring systems are deployed in locations where regular battery replacement or charging is not feasible. Since these sensors often operate in remote areas, powering them efficiently is crucial for long-term sustainability. Traditional batteries can be costly, environmentally damaging, and require frequent maintenance, making power-efficient solutions essential.
- **Solution**: Energy harvesting technologies, such as solar, wind, or kinetic-powered sensors, offer a potential solution to mitigate power consumption. However, maintaining a reliable

power source for continuous operation remains challenging, especially for large-scale or long-term deployments.

- **Impact**: Power limitations can significantly reduce the lifespan and efficiency of IoT sensors, necessitating the development of more energy-efficient hardware and energy harvesting technologies.

6.4.3 Data Overload: Managing the Massive Volume of Data

- **Challenge**: IoT systems generate vast amounts of real-time environmental data from numerous sensors deployed across large areas. This can result in data overload, overwhelming traditional data processing systems that may not be equipped to handle such large volumes of information. Storing and processing data from thousands or millions of sensors require substantial computational resources and can result in delays or inefficiencies.

- **Solution**: To address data overload, edge computing, where data are processed closer to the source (i.e., at the sensor level or on localized servers), can help reduce the volume of data transmitted to central storage systems. Data filtering and compression techniques can also minimize bandwidth usage. Additionally, AI and Machine Learning algorithms can analyze data in real time, extracting meaningful insights and reducing the need for storing excessive raw data.

- **Impact**: Efficient data management solutions are critical to ensure the effective use of IoT systems in environmental monitoring, preventing bottlenecks and ensuring timely insights.

6.4.4 Security and Privacy: Safeguarding Sensitive Data

- **Challenge**: IoT devices deployed for environmental monitoring often collect sensitive data related to public health, such as air quality levels, water contamination, or hazardous substances in soil. These data can have significant implications for public safety and requires strong security measures. IoT systems are vulnerable to cyberattacks, including data breaches, unauthorized access, and tampering with the collected data.

- **Solution**: Ensuring robust data security through encryption, secure communication protocols, and multi-layered authentication mechanisms is essential to safeguard the integrity and privacy of environmental data. IoT devices should be designed with built-in security features to minimize vulnerabilities.
- **Impact**: Without proper security measures, IoT systems in environmental applications may risk data breaches or manipulation, compromising the accuracy and trustworthiness of environmental monitoring.

6.4.5 Interoperability: Ensuring Device Compatibility

- **Challenge**: In large-scale environmental monitoring systems, IoT devices from various manufacturers often need to communicate and share data seamlessly. However, differences in communication protocols, data formats, and system architectures can create interoperability issues.
- **Solution**: Standardizing communication protocols, data formats, and APIs can ensure compatibility across different IoT devices, regardless of the manufacturer. Open-source frameworks and collaborative efforts in the IoT ecosystem are key to achieving interoperability and enabling the integration of diverse sensor types.
- **Impact**: Without interoperability, the integration of various devices may become cumbersome, leading to inefficiencies and increased complexity in environmental monitoring systems.

6.4.6 Cost and Scalability: Balancing Affordability and Expansion

- **Challenge**: The deployment and maintenance of a large number of IoT sensors across vast geographical areas can be expensive, especially in environmental applications where long-term monitoring is required. The cost of sensors, infrastructure, data storage, and processing can quickly add up, making large-scale deployments unaffordable for many organizations or governments.

- **Solution**: Cost-effective solutions, such as low-cost sensors, scalable cloud storage, and efficient data processing algorithms, are essential to reduce deployment costs. Partnerships with governments or organizations for funding, as well as leveraging open-source technologies, can also help alleviate financial barriers.

- **Impact**: Without cost-effective solutions, the scalability of IoT-based environmental monitoring systems may be limited, preventing widespread implementation and reducing their overall impact.

While the potential of IoT in environmental applications is vast, addressing these challenges is essential to ensure the successful deployment and operation of such systems. Ongoing advancements in IoT technology, including improved communication protocols, power management techniques, data processing capabilities, and security measures, continue to enhance the efficiency and scalability of environmental monitoring systems. As these issues are tackled, IoT can play a pivotal role in transforming how we monitor, manage, and protect the environment on a global scale.

7
AI TECHNIQUES FOR ENVIRONMENTAL DATA ANALYSIS

7.1 Introduction to Machine Learning and Deep Learning

Artificial Intelligence (AI) has become an indispensable tool in environmental science (Figure 7.1, revolutionizing the way environmental data are analyzed and applied (Cui et al., 2023). By leveraging advanced computational techniques, AI enables researchers and policymakers to transform vast amounts of raw, often complex, environmental data into actionable insights, paving the way for informed and effective decision-making. Two primary AI techniques that stand out for their application in environmental data analysis are Machine Learning (ML) and deep learning (DL).

7.2 Machine Learning (ML)

ML is a core branch of AI that focuses on creating systems capable of learning from data and improving their performance over time

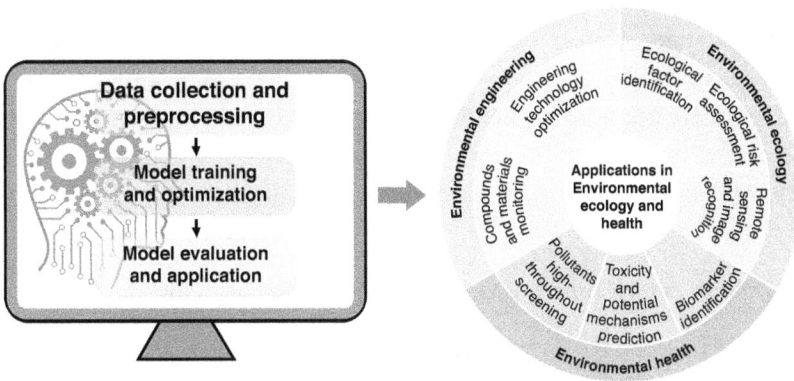

Figure 7.1 Framework for applying AI techniques in environmental ecology and health (Cui et al., 2023).

DOI: 10.1201/9781003712701-9

without explicit programming for every possible scenario. In environmental monitoring, ML offers robust methods to analyze large and often disparate datasets obtained from diverse sources, including ground-based sensors, satellite imagery, weather stations, and Internet of Things devices.

By identifying patterns and trends, ML algorithms can perform tasks such as classifying land use types, predicting air quality levels, detecting anomalies in climate data, and forecasting weather or pollution events. For instance, in air quality management, ML models can predict pollutant dispersion patterns based on meteorological data, aiding in devising strategies to mitigate air pollution exposure. Similarly, in agriculture, ML algorithms can classify crop health based on hyperspectral data, enabling precision farming practices.

7.3 Deep Learning (DL)

DL, a specialized and powerful subset of ML, employs neural networks with multiple layers to analyze and model complex and high-dimensional relationships within data. The term "deep" refers to the extensive depth of layers in these networks, which allows them to extract hierarchical features and capture intricate patterns that are often challenging for traditional ML methods. DL techniques are especially effective for tasks involving large-scale and unstructured datasets, such as high-resolution satellite images, videos, and textual documents. In environmental applications, DL excels in tasks like remote sensing image analysis, where it can identify and quantify land cover changes, detect deforestation or urban expansion, and assess disaster-impacted regions with unparalleled accuracy. Furthermore, DL models are used to analyze unstructured data such as environmental reports, enabling the extraction of insights about trends in policy changes or environmental regulations.

7.4 Synergy of ML and DL in Environmental Applications

Both ML and DL are inherently data-driven, relying on historical datasets to build predictive models. Once trained, these models are capable of analyzing new data in real-time, offering predictive insights and aiding decision-making processes. ML techniques are often used for structured data analysis, while DL models are ideal for handling unstructured and complex data, making them complementary in environmental data analysis workflows.

For example, hybrid AI systems that integrate ML and DL can detect anomalies in air quality datasets while also analyzing remote sensing imagery to identify pollution sources. Such integrated approaches enhance the ability to address multifaceted environmental challenges, ranging from climate change impact assessments to biodiversity conservation.

7.5 Contribution to Sustainable Environmental Management

The adoption of ML and DL techniques in environmental data analysis has significantly improved the precision, scalability, and speed of monitoring and forecasting systems. By automating the analysis of vast datasets, these techniques enable timely identification of environmental issues, support evidence-based policymaking, and facilitate proactive management strategies. For instance, early detection of extreme weather events through AI-powered systems can help minimize damage and protect vulnerable communities.

In conclusion, the use of AI techniques like ML and DL in environmental data analysis is revolutionizing the field, offering innovative solutions to complex problems while contributing to the broader goals of sustainability and environmental stewardship. As these technologies continue to evolve, their integration into environmental science will undoubtedly expand, further enhancing our ability to monitor, protect, and manage the natural world.

7.6 Algorithms for Classification, Prediction, and Clustering

Several AI algorithms are used in environmental data analysis (Figure 7.2), each serving a specific purpose depending on the type of analysis required (Minh et al., 2021). Common AI techniques include:

- **Classification Algorithms**: Classification is used when the goal is to categorize data into predefined groups. For example, a classification algorithm can be used to categorize air quality

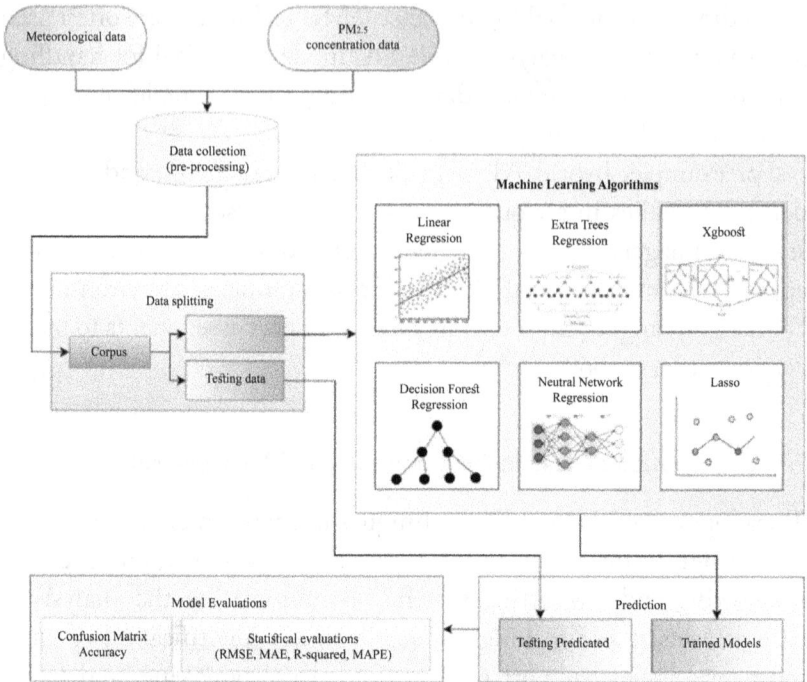

Figure 7.2 Machine Learning flow diagram using six algorithms (Minh et al., 2021).

data into different pollution levels (e.g., low, medium, high). Common classification algorithms include:

- **Decision Trees**: These algorithms split data into branches based on feature values to classify data into categories.
- **Random Forests**: An ensemble of decision trees that improves accuracy and reduces overfitting by combining the results of multiple trees.
- **Support Vector Machines (SVM)**: SVM separates data into classes by finding the optimal hyperplane that maximizes the margin between classes.
- **k-Nearest Neighbours (k-NN)**: A simple algorithm that classifies data based on the majority class of the nearest data points.
- **Prediction Algorithms**: Prediction is used when the goal is to forecast future values based on historical data. In environmental monitoring, prediction algorithms can be used to

predict air quality levels, temperature, or pollutant concentrations. Common prediction algorithms include:

- **Linear Regression**: A basic technique that models the relationship between a dependent variable and one or more independent variables. It is used for predicting continuous values, such as predicting temperature changes.
- **Time Series Analysis**: Algorithms like Autoregressive Integrated Moving Average are used to forecast future values based on past observations, commonly applied in predicting air quality trends over time.
- **Neural Networks**: DL models, such as feed-forward neural networks, can predict outcomes based on complex, non-linear relationships in environmental data.

- **Clustering Algorithms**: Clustering is used to group similar data points together based on their features. In environmental data analysis, clustering can identify patterns in pollutant distribution, habitat types, or species populations. Common clustering algorithms include:
  - **k-Means Clustering**: A widely used algorithm that partitions data into k clusters by minimizing the distance between data points and the centroid of each cluster.
  - **DBSCAN (Density-Based Spatial Clustering of Applications with Noise)**: A clustering algorithm that groups points based on their density, ideal for identifying regions with high concentrations of pollutants in air or water.
  - **Hierarchical Clustering**: Builds a tree of clusters that can be used for exploring the relationships between different groups, such as areas with similar environmental conditions.

These algorithms are essential tools in the analysis of environmental data, enabling AI systems to classify data, predict trends, and discover hidden patterns in complex datasets.

7.7 Case Studies: Forecasting Pollution and Environmental Trends

AI techniques have been successfully applied in several case studies related to forecasting pollution and analyzing environmental trends. These case studies demonstrate how ML and DL are used to enhance environmental monitoring and decision-making.

- **Air Pollution Forecasting Using ML**: In a study on air quality prediction, ML algorithms like Random Forests and Support Vector Machines were used to forecast air pollution levels in urban areas. The models were trained on historical data collected from air quality sensors and meteorological stations, and they successfully predicted PM2.5 and PM10 concentrations, which are critical for public health assessments. The study showed how ML can provide timely air quality predictions that can help policymakers issue early warnings and implement preventive measures.

- **Water Quality Prediction Using Neural Networks**: In a case study of water quality monitoring in a river system, DL algorithms were used to predict parameters such as pH, dissolved oxygen, and levels of contaminants (e.g., nitrates and heavy metals). Recurrent neural networks were applied to time-series data, allowing for accurate predictions of water quality trends based on historical data. This prediction system enables better management of water resources and helps prevent pollution-related issues before they escalate.

- **Forest Fire Risk Prediction Using AI**: AI has also been applied in predicting forest fire risks by analyzing environmental data such as temperature, humidity, wind speed, and vegetation density. ML algorithms, including decision trees and neural networks, were used to classify regions with high fire risk based on these parameters. The predictions help authorities to implement fire prevention strategies in high-risk areas, potentially saving lives and resources.

- **Climate Change and Temperature Prediction**: AI techniques, particularly time series analysis and neural networks, have been used to forecast temperature and precipitation trends in the context of climate change. These models are based on historical data and predict future climate variables, which are essential for policy planning and agricultural management. AI-based systems can help assess the impact of climate change on ecosystems and human societies, allowing for more proactive and adaptive measures.

These case studies illustrate the power of AI in forecasting pollution, monitoring environmental changes, and providing data-driven solutions for sustainable resource management and environmental protection. The integration of ML and DL into environmental monitoring systems allows for more accurate predictions, faster response times, and better-informed decision-making.

8

DATA MANAGEMENT AND ANALYTICS

8.1 Big Data in Environmental Monitoring

In environmental monitoring, the volume, variety, and velocity of data collected from IoT sensors, satellite imagery, weather stations, and other sources contribute to the phenomenon known as "Big Data." Big Data analytics play a crucial role in managing and extracting insights from this vast amount of data. Key characteristics of Big Data in environmental monitoring include:

- **Volume:** Environmental monitoring systems generate large volumes of data. Sensors deployed in urban areas, forests, oceans, and agricultural fields continuously collect data on air quality, water levels, soil conditions, weather, and more. This results in an overwhelming amount of data that requires specialized storage and processing systems.
- **Velocity:** Environmental data are often time-sensitive. For example, air quality and weather data require real-time monitoring to enable timely interventions. The speed at which data are collected, processed, and analyzed is essential for effective decision-making, particularly in dynamic environments like pollution forecasting, flood monitoring, and disaster management.
- **Variety:** Environmental data comes in many forms: structured data (e.g., numerical readings from sensors), unstructured data (e.g., satellite images), and semi-structured data (e.g., text data from environmental reports). Big Data techniques enable the integration and analysis of this diverse range of data to provide comprehensive insights into environmental conditions.

DOI: 10.1201/9781003712701-10

The ability to harness Big Data analytics enables environmental monitoring systems to detect patterns, make predictions, and optimize resource management. By applying AI and IoT to Big Data, decision-makers can gain a deeper understanding of complex environmental phenomena, identify emerging trends, and implement proactive measures to protect the environment.

8.2 Data Cleaning, Storage, and Integration

Given the sheer volume and variety of environmental data, proper data management techniques are essential for ensuring its quality and usability. Key processes in data management include:

- **Data Cleaning**: Raw environmental data often contains noise, errors, and inconsistencies, such as missing values, duplicate entries, or outliers. Data cleaning involves identifying and rectifying these issues to ensure the data is accurate and reliable. Techniques such as outlier detection, imputation of missing values, and filtering of irrelevant data are employed to prepare data for analysis. Clean data are critical for Machine Learning models, as erroneous or incomplete data can lead to biased results or inaccurate predictions.

- **Data Storage**: Environmental data storage must accommodate the large volumes of data generated by IoT sensors and other monitoring tools. Cloud storage solutions, such as Amazon Web Services, Google Cloud, and Microsoft Azure, offer scalable, cost-effective storage options for Big Data. Additionally, specialized databases like NoSQL databases (e.g., MongoDB, Cassandra) are used to store unstructured or semi-structured data, while traditional relational databases (e.g., SQL) handle structured data. A well-organized data storage system ensures easy access, retrieval, and scalability for future data analysis.

- **Data Integration**: Environmental data are often collected from multiple sources, such as different sensors, remote sensing platforms, and historical datasets. Integrating this diverse data is essential for creating a unified view of the environmental conditions being monitored. Data integration techniques involve combining data from heterogeneous sources and ensuring that the data are aligned in terms of time, format, and context. Data fusion techniques are used to merge data from IoT sensors with satellite imagery, weather forecasts, and historical environmental data to enhance the comprehensiveness and accuracy of environmental models.

Effective data integration provides a holistic view of environmental conditions and enables more informed decision-making by providing a clear picture of the state of the environment across different scales and locations.

8.3 Visualization Techniques for Environmental Data

Data visualization plays a vital role in interpreting complex environmental data and communicating insights effectively. Visualizations help decision-makers, scientists, and the public to better understand trends, patterns, and anomalies in environmental data. Key visualization techniques for environmental data include:

- **Geospatial Visualization**: Many environmental data sources, such as satellite imagery, weather stations, and IoT sensors, involve spatial information. Geographic information systems (GIS) are used to create maps that display data geographically. For example, air quality monitoring data can be visualized on a map, with color-coded areas indicating pollutant concentrations. GIS maps also allow for the integration of multiple data layers (e.g., temperature, humidity, pollution levels), providing a comprehensive spatial view of environmental conditions.
- **Time Series Visualization**: Time series data, such as temperature fluctuations or pollutant levels over time, can be visualized using line graphs, bar charts, and histograms. These visualizations help to identify trends, seasonal variations, and sudden spikes or drops in environmental parameters. Time series analysis allows for forecasting future trends based on historical data, which can be crucial for environmental planning and policy-making.

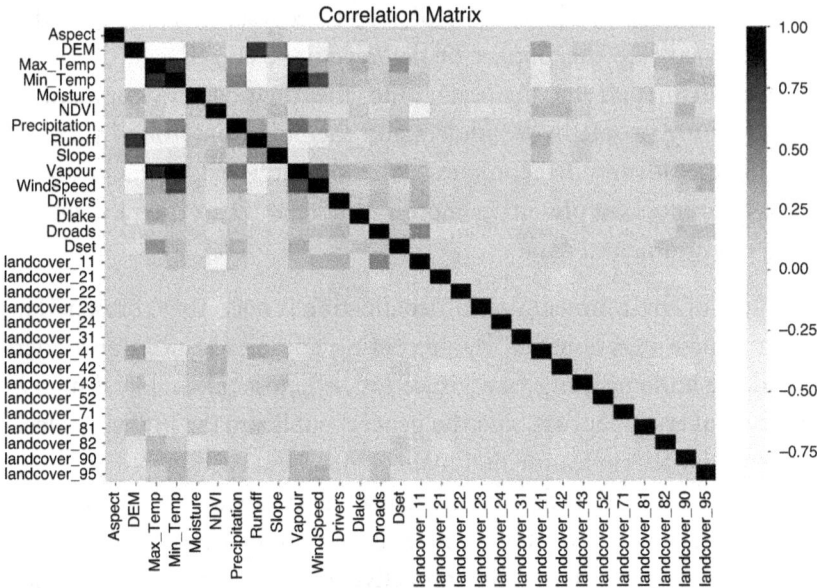

Figure 8.1   Heat map of the factors used for predicting forest fires.

- **Heatmaps**: Heatmaps are used to visualize the intensity of environmental parameters across a given area (Figure 8.1). For example, heatmaps can display variations in air pollution concentrations or temperature distribution over a geographical area, allowing for quick identification of hotspots. These visualizations are particularly useful in urban air quality monitoring and disaster management, where spatial patterns are crucial for identifying areas of concern.

- **Dashboards**: Interactive dashboards are used to display real-time environmental data from multiple sources. These dashboards can include various visualizations, such as graphs, maps, and tables, to provide a comprehensive overview of environmental conditions. Dashboards can be customized for different user needs, from government agencies monitoring air quality to researchers tracking water pollution in real time. Dashboards also allow for the integration of predictive analytics, where future trends can be visualized alongside current data.

- **3D Visualizations and Augmented Reality (AR)**: In advanced applications, 3D visualizations and AR can be used to display environmental data in an immersive manner.

For example, 3D models of forest ecosystems can be used to visualize the impact of deforestation, or AR can overlay real-time air quality data onto the physical environment, helping people better understand the pollution around them. These advanced techniques enhance public engagement and awareness by providing more intuitive ways to interact with environmental data.

The goal of environmental data visualization is not only to present data but to make it actionable. By presenting complex data in an understandable and engaging way, visualization helps stakeholders, such as policy makers, scientists, and the general public, make informed decisions for better environmental management and sustainability.

Through effective data management and advanced visualization techniques, environmental monitoring systems can turn raw data into valuable insights, facilitating proactive responses to environmental challenges and supporting sustainable practices across various sectors.

9

GLOBAL CASE STUDIES FROM UNDERREPRESENTED REGIONS

9.1 Africa and Middle East

- **Ghana—Google Project Relate**: Originally created to assist individuals with speech impairments, this voice recognition tool is now piloted in Ghana. It supports people with conditions like cleft palate to communicate clearly, boosting their confidence and employability.
- **South Africa and Uganda—AI for Tuberculosis (TB) Screening**: In South Africa, AI-driven tools help speed up TB diagnosis. In Uganda, a portable digital X-ray backpack with integrated AI has screened over 50,000 people and diagnosed more than 1,000 TB cases.
- **Ethiopia—Speech-to-Speech Translator**: Developed under Grand Challenges Ethiopia, this tool enables multilingual communication between patients and doctors by translating between Amharic, Oromo, Somali, and English.

9.2 Asia and Pacific

- **SERVIR Mekong**: A NASA-USAID initiative uses satellite imagery and AI to monitor flooding, land use, and ecosystem changes in Cambodia, Laos, Myanmar, Thailand, and Vietnam.
- **Sri Lanka—Participatory Large Language Model (LLM) Agents**: A multilingual LLM system facilitates community-led research and inclusive decision-making in Sinhala-speaking areas, particularly those vulnerable to environmental stress.

DOI: 10.1201/9781003712701-11

9.3 Latin America and Caribbean

- **AIME (Dominican Republic, Brazil, Philippines)**: Founded by Rainier Mallol, AIME's AI platform forecasts vector-borne diseases (like dengue and Zika) with up to 87% accuracy, showcasing strong South-South collaboration.
- **Colombia—AI4PEP Network**: Part of a 16-country Global South initiative, AI4PEP uses AI to strengthen pandemic preparedness and improve disease surveillance and public health responses.
- **Latam GPT (Launching September 2025)**: Developed by CENIA (Chile) and over 30 partner institutions, this open-source LLM is multilingual and includes Indigenous languages like Rapa Nui. It is designed for use in public services, education, and healthcare.
- **Honduras—Informal Settlement Mapping in Tegucigalpa**: Machine Learning and satellite data are used to map slums with high precision, helping NGOs plan surveys and distribute resources more effectively.

PART III
APPLICATIONS AND PRACTICAL INSIGHTS

10

AIR QUALITY
MONITORING SYSTEMS

10.1 AI-Driven Models for Air Quality Index Prediction

As air pollution continues to rise globally, air quality monitoring has become a cornerstone of efforts to protect public health and inform environmental policies. Predicting air quality levels accurately is essential for mitigating the adverse impacts of air pollution on human health and ecosystems. Leveraging Artificial Intelligence (AI) in air quality monitoring systems is revolutionizing the field, providing real-time, accurate, and dynamic forecasting capabilities that surpass traditional approaches.

10.2 AI-Driven Models for Air Quality Index Prediction

AI models, powered by Machine Learning (ML) and deep learning (DL), are transforming the prediction and management of the air quality index (AQI), offering significant improvements in accuracy and responsiveness.

10.3 Machine Learning Models

ML algorithms such as Random Forests, support vector machines, and neural networks are widely employed to forecast AQI based on historical and real-time data. These models analyze and interpret patterns in air pollutant concentrations—such as PM2.5, PM10, CO, NOx, and SO_2—in conjunction with weather variables like temperature, humidity, and wind speed.

By training on large, diverse datasets, these models can predict AQI levels with high reliability, enabling authorities to issue timely air quality alerts. For instance, ML models can detect spikes in pollution

DOI: 10.1201/9781003712701-13

due to localized events such as traffic congestion or industrial emissions, providing actionable insights for interventions. Additionally, these algorithms can be tailored to specific geographical regions, accounting for local environmental and meteorological factors.

10.4 Deep Learning for Real-Time Prediction

DL techniques, including recurrent neural networks (RNNs) and long short-term memory (LSTM) networks (Figure 10.1; Kim et al., 2022), have proven particularly effective in handling the temporal and sequential nature of air pollution data.

Unlike traditional models, which may struggle with non-linear relationships or time dependencies, RNNs and LSTMs excel at capturing these dynamics, making them ideal for real-time AQI prediction. These models can analyze and predict pollution trends over time, identifying cyclical patterns, such as daily traffic or seasonal variations, that impact air quality. For example, during peak traffic hours, LSTMs can provide real-time predictions of AQI, allowing cities to implement short-term mitigation measures like traffic rerouting or public transportation advisories.

10.5 Air Quality Forecasting with Big Data

The integration of AI with Big Data analytics has further enhanced the predictive capabilities of air quality monitoring systems. AI-driven

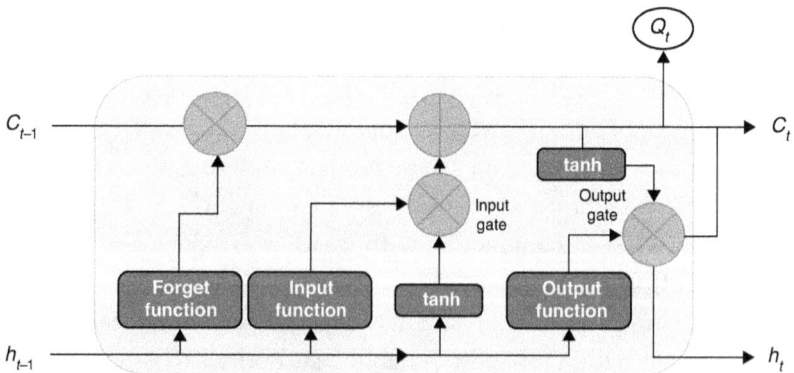

Figure 10.1 Conceptual diagram of the LSTM model (Kim et al., 2022).

models now utilize vast amounts of data collected from diverse sources, including:

- **Internet of Things (IoT) sensors** installed across cities for localized air quality readings.
- **Satellite imagery** for tracking regional and global pollution trends.
- **Weather stations** for real-time meteorological inputs.
- **Traffic monitoring systems** to evaluate emissions from vehicles in urban areas.

By synthesizing these inputs, AI models can provide a more comprehensive understanding of how different factors—such as industrial emissions, vehicle density, or meteorological changes—contribute to air pollution. This holistic perspective allows for multi-factorial analysis and predictions, enabling policymakers to design proactive strategies for managing air quality.

10.6 Impact of AI-Driven Air Quality Systems

AI-powered air quality monitoring systems have broad applications in urban planning, public health, and environmental management:

- **Urban Planning**: AI models can guide city planners by identifying high-risk areas and recommending solutions such as green buffer zones or traffic control measures.
- **Public Health Alerts**: By predicting AQI levels, AI systems enable governments to issue warnings, helping individuals reduce exposure during hazardous conditions.
- **Industrial Emission Control**: Real-time monitoring allows industries to adjust processes to meet regulatory standards, reducing their environmental footprint.

In summary, the integration of AI-driven models into air quality monitoring systems represents a paradigm shift, enabling data-driven, real-time responses to air pollution challenges. These advancements not only improve the accuracy and timeliness of air quality predictions but also empower communities and governments to take proactive measures, fostering healthier and more sustainable environments.

10.7 IoT Networks for Real-Time Air Monitoring

IoT has significantly transformed air quality monitoring by enabling continuous, real-time data collection and transmission (Figures 10.2 and 10.3; Minh et al., 2021; Ahmad & Ahmad, 2023). Through IoT networks, environmental data are gathered and shared instantly, providing up-to-date information on air quality across large areas. This technological advancement allows for more precise monitoring and timely responses to air pollution, ultimately improving public health and urban planning efforts.

10.8 IoT Sensors

At the heart of IoT-based air quality monitoring systems are small, low-cost sensors that can be deployed in various locations, such as residential neighborhoods, schools, industrial zones, and transportation hubs. These sensors are designed to measure key air pollutants,

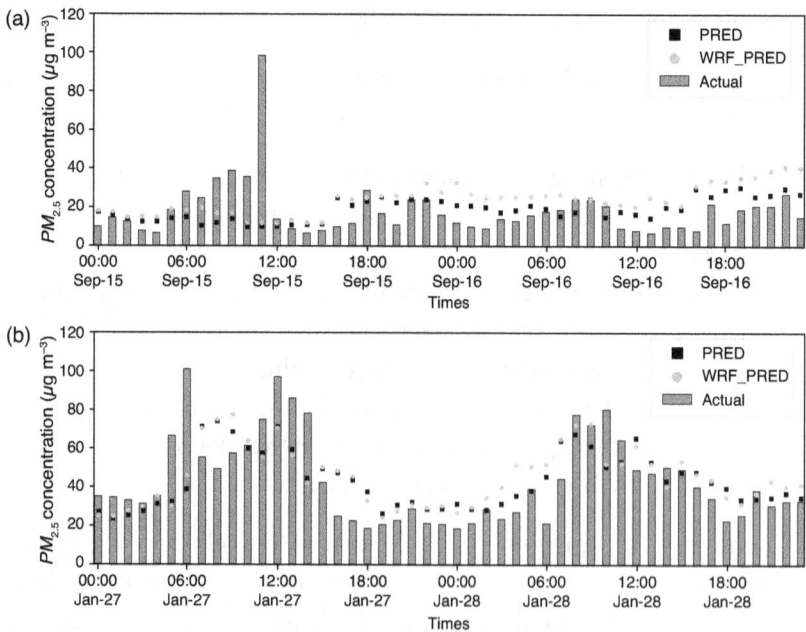

Figure 10.2 Actual and predicted PM2.5 concentration with observation and simulated meteorological data; (a) in the rainy season (September 2020) and (b) in the dry season (January 2021) (Minh et al., 2021).

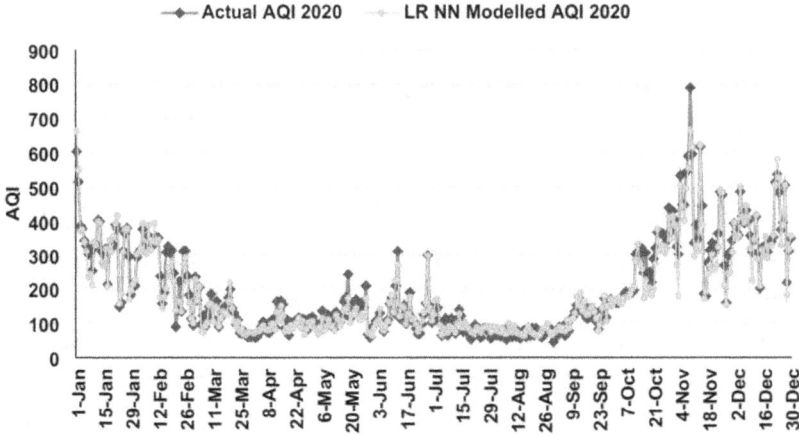

Figure 10.3 Average daily AQI predicted by LR-NN model and actual AQI for the year 2020 (Ahmad & Ahmad, 2023).

including particulate matter (PM2.5, PM10), nitrogen dioxide (NO_2), sulfur dioxide (SO_2), carbon monoxide (CO), ozone (O_3), and volatile organic compounds (VOCs).

The sensors are equipped with wireless communication capabilities, which enable them to transmit the collected data in real-time to central servers or cloud platforms for further processing and analysis. The low cost and small size of these sensors make them highly scalable, allowing for the creation of extensive air quality monitoring networks that can cover wide geographical areas. By placing these sensors in strategic locations, cities and organizations can obtain a detailed, real-time picture of air pollution levels in specific regions or across broader urban landscapes.

10.9 Network Architecture

The efficiency of an IoT-based air quality monitoring system is dependent on its network architecture. Most systems utilize mesh or star network configurations to facilitate smooth communication between distributed sensors and central data processing platforms.

- **Mesh Networks**: In a mesh network, each sensor acts as a node that can communicate with multiple other sensors, creating a robust, decentralized system. This configuration ensures that if one sensor fails or is offline, others can still relay data, making the system highly resilient.

- **Star Networks**: In star networks, sensors communicate directly with a central hub or gateway. This centralized model can simplify data management but may be more vulnerable to failures if the central node experiences issues.

Communication protocols used in IoT-based air quality monitoring include Wi-Fi, low power wide area network (LPWAN), and cellular networks. These protocols ensure efficient data transmission over various distances, allowing real-time updates even in remote or rural areas. Once the data are collected from the sensors, the data are transmitted to centralized platforms, where it can be processed, analyzed, and visualized for actionable insights.

10.10 Real-Time Monitoring and Alerts

One of the key advantages of IoT-based air quality monitoring systems is the ability to perform real-time monitoring of air quality. These systems provide continuous updates on pollution levels, offering up-to-the-minute data that can be used to assess air quality at any given moment. When air pollutant concentrations exceed predefined thresholds, IoT systems can trigger alerts using integrated AI algorithms. These alerts can be sent to various stakeholders, including government agencies, public health organizations, and the general public. For example, if PM2.5 levels rise to hazardous levels, the system can issue warnings, advising individuals with respiratory conditions to avoid outdoor activities. Furthermore, IoT systems allow for localized air quality information, meaning individuals can access real-time data specific to their location. This empowers people to make informed decisions about their exposure to pollution, whether it be altering their outdoor plans or taking protective measures such as wearing masks.

10.11 Integration with Smart City Infrastructure

Another major advantage of IoT-based air quality monitoring is its ability to integrate with other components of smart city infrastructure. For example, when air quality data are combined with smart traffic management systems, cities can better understand how traffic congestion contributes to pollution levels. By analyzing these data, authorities can implement strategies to reduce emissions, such as:

- **Optimizing traffic signals** to improve traffic flow and reduce idling time for vehicles, which in turn decreases air pollution.
- **Encouraging the use of electric vehicles (EVs)** by providing incentives or creating more EV-friendly infrastructure.

Additionally, IoT-based air quality systems can be integrated with energy-efficient building technologies, where data on indoor air quality can be used to optimize HVAC systems, improve ventilation, and ensure that buildings maintain healthy air standards. This synergy between air quality monitoring and other smart city functions creates a more holistic and sustainable approach to managing urban environments.

10.12 Practical Examples: Urban and Industrial Air Quality Studies

AI and IoT-powered air quality monitoring systems have been successfully implemented in both urban and industrial settings, offering substantial improvements in tracking pollution levels and assessing their impacts on public health and the environment. The integration of these advanced technologies is revolutionizing the way cities and industries approach air quality management, enabling more proactive and data-driven decision-making. Below are some key practical examples of these systems in action:

10.13 Urban Air Quality Monitoring in Smart Cities

In cities such as Beijing, London, and New York, advanced air quality monitoring systems are actively being deployed to provide real-time data on urban pollution levels (Figure 10.4; Du et al., 2017). These cities use a combination of IoT sensors and AI-driven models to track key pollutants like PM2.5, NOx, and ozone, offering a comprehensive understanding of air quality across different parts of the city.

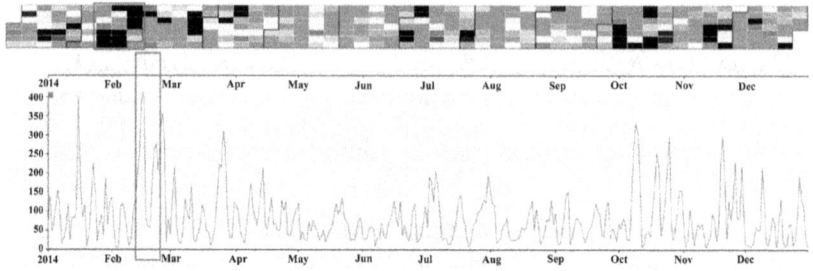

Figure 10.4 The daily trends of PM2.5 in Beijing (Du et al., 2017).

- New York has integrated a network of air quality sensors into its broader smart city infrastructure. The sensors collect real-time data on pollutant concentrations, and the data are made publicly accessible through mobile apps and online platforms. This allows residents to monitor the air quality in their neighborhoods, helping them make informed decisions, such as adjusting outdoor activities or taking preventive health measures when pollution levels rise.
- In London, similar systems are in place, where data from air quality monitoring sensors are analyzed to track trends and predict future pollution levels. This information helps city officials implement measures like reducing traffic emissions, issuing air quality alerts, and informing residents about when to take actions to protect their health.

These urban air quality systems are key components of smart city development, making use of advanced sensors, data analytics, and real-time predictions to improve urban living conditions and public health.

10.14 Industrial Air Quality Monitoring

In industrial settings, air quality monitoring plays a critical role in ensuring compliance with environmental regulations and safeguarding workers' health. Manufacturing plants, power plants, and refineries often produce large quantities of air pollutants, including SO_2, nitrogen oxides (NO_2), and particulate matter. Monitoring these emissions is vital for preventing environmental degradation and adhering to legal standards.

- IoT sensors are installed at various points in industrial facilities to continuously measure pollutant levels. These sensors are integrated with AI-driven systems to analyze the data and provide real-time insights on emissions. AI algorithms can also predict emission trends, allowing operators to optimize industrial processes, reduce pollution, and maintain air quality within regulatory limits.
- For example, power plants equipped with IoT sensors can detect and predict SO_2 emissions, enabling the plant to adjust operations to reduce the release of harmful pollutants into the atmosphere. This proactive approach ensures that emissions remain within permissible thresholds, preventing costly fines and contributing to environmental protection.

10.15 Smart Monitoring for Environmental Protection

In regions with significant agricultural activity, particularly in countries like India and China, the use of AI and IoT systems is helping to monitor the impact of pesticides and fertilizers on air quality. These chemicals, when sprayed in large quantities, can release harmful compounds such as ammonia and VOCs into the air, which can degrade air quality and harm both human health and the environment.

- IoT sensors are placed near agricultural zones to monitor the concentrations of ammonia, VOCs, and other pollutants. These data are then analyzed by AI models to predict the risk of air pollution caused by agricultural practices.
- By integrating these monitoring systems into agricultural practices, policymakers and farmers can make informed decisions about when and how to apply pesticides and fertilizers, reducing the impact of these activities on air quality. AI systems also enable the prediction of pollution spikes, helping to implement sustainable farming practices and minimize the environmental footprint of agriculture.

This approach not only promotes healthier air but also encourages the adoption of sustainable farming practices, benefiting both the environment and the agricultural community.

10.16 Collaborative Environmental Monitoring Networks

In some cities, public and private entities have come together to build large-scale air quality monitoring networks that involve the community in data collection. These collaborative networks enhance the spatial coverage of air quality data, empowering citizens to actively participate in environmental protection efforts.

- For example, **AQI** networks in various cities enable residents to install low-cost air quality sensors in their homes or local communities. These sensors collect air quality data, which are then shared with central platforms.
- By integrating these data, the network provides a comprehensive and more granular view of air quality across different neighborhoods, allowing local authorities to track pollution trends in real-time. In return, the public is empowered to take action, whether it's adjusting daily routines based on air quality data or advocating for local policy changes to improve air quality.

These collaborative efforts help to democratize air quality monitoring, engaging the broader community and creating a sense of collective responsibility for improving environmental conditions.

11

WATER QUALITY MONITORING SYSTEMS

11.1 IoT-Based Solutions for Water Contamination Detection

Water quality monitoring is essential for safeguarding public health, preserving aquatic ecosystems, and efficiently managing water resources. Traditional methods of water quality assessment often involve sporadic sampling and laboratory-based analysis, which can be both time-consuming and costly. In contrast, Internet of Things (IoT)-based solutions are revolutionizing water quality monitoring by enabling continuous, real-time data collection, offering a more dynamic and proactive approach to detecting water contamination.

11.2 IoT Sensors for Water Quality

IoT sensors are deployed in various water bodies, including rivers, lakes, reservoirs, and groundwater wells, to continuously monitor and measure a range of critical water quality parameters. These sensors are capable of detecting factors that directly impact water safety, such as:

- **pH levels**: This indicates the acidity or alkalinity of water, which is vital for aquatic life and safe drinking water.
- **Dissolved Oxygen (DO)**: Essential for supporting aquatic organisms, a decrease in DO levels can indicate contamination or eutrophication.
- **Turbidity**: The cloudiness or haziness of water, which can be caused by pollutants or suspended particles.
- **Temperature**: Water temperature influences the solubility of oxygen and the health of aquatic life.
- **Salinity**: Important in monitoring water bodies, especially in coastal or estuarine areas where saltwater intrusion may affect freshwater resources.

DOI: 10.1201/9781003712701-14

- **Pollutants**: Sensors can detect hazardous substances such as nitrates, heavy metals, and pesticides that could harm both ecosystems and human health.

These IoT sensors are equipped with wireless communication technology, enabling them to transmit collected data in real-time to cloud-based platforms. The data are then processed and analyzed, allowing for the continuous monitoring of water quality across large and diverse water bodies.

11.3 Remote Water Monitoring

One of the key advantages of IoT-based water quality monitoring is the ability to operate in remote or difficult-to-access locations. Many water bodies, particularly in rural, protected, or undeveloped areas, may not have the infrastructure for frequent sampling or on-site testing. IoT-based systems overcome this limitation by using low-power communication networks, such as low power wide area networks, to transmit data from sensors to centralized systems.

These remote water monitoring systems eliminate the need for manual sampling, thus making the monitoring process more cost-effective and efficient. The ability to place sensors in remote locations ensures that even water bodies that are far from urban centers, or in protected zones where access is limited, can be continuously monitored for any potential water quality issues. This continuous stream of data provides a comprehensive and up-to-date picture of water conditions, which can be crucial for timely interventions.

11.4 Early Detection of Water Contamination

IoT-based water quality monitoring systems are particularly effective in providing early detection of potential water contamination events. This capability is vital for mitigating risks before contamination spreads or becomes a serious threat to public health and the environment. Some key early warning functions include:

- **Chemical Spills**: IoT sensors can detect sudden spikes in specific pollutants, such as heavy metals or pesticides, signaling a chemical spill or industrial discharge into the water.
- **Microbial Contamination**: Changes in water quality, such as a sudden drop in DO levels or an increase in turbidity, can indicate the presence of harmful microbial growth, often as a result of sewage or organic pollution.
- **Nutrient Levels**: An increase in nutrient levels, particularly nitrates or phosphates, can lead to eutrophication, where excessive nutrients promote algae blooms that deplete oxygen levels and harm aquatic life.

By monitoring critical parameters continuously, IoT systems are capable of identifying anomalies in real time, such as an unexpected change in turbidity or a rapid drop in oxygen levels. These anomalies serve as early indicators of contamination, allowing for immediate responses, such as:

- **Alerting local authorities** to potential contamination, enabling them to take swift action such as issuing warnings, closing water sources, or initiating clean-up efforts.
- **Providing data to water treatment plants,** allowing them to adjust filtration and treatment processes in response to detected contamination, ensuring safe drinking water for communities.

11.5 Predictive Models for Water Quality Trends

As environmental challenges related to water quality continue to grow, Artificial Intelligence (AI) and Machine Learning (ML) are becoming indispensable tools for predicting water quality trends. By leveraging historical and real-time data, predictive models enable better management of water resources and the anticipation of pollution events before they reach critical levels. These predictive capabilities help policymakers, environmental agencies, and water management authorities make informed decisions and take proactive actions to maintain and improve water quality. A prediction model for water consumption has been given in Figure 11.1 (Rustam et al., 2022).

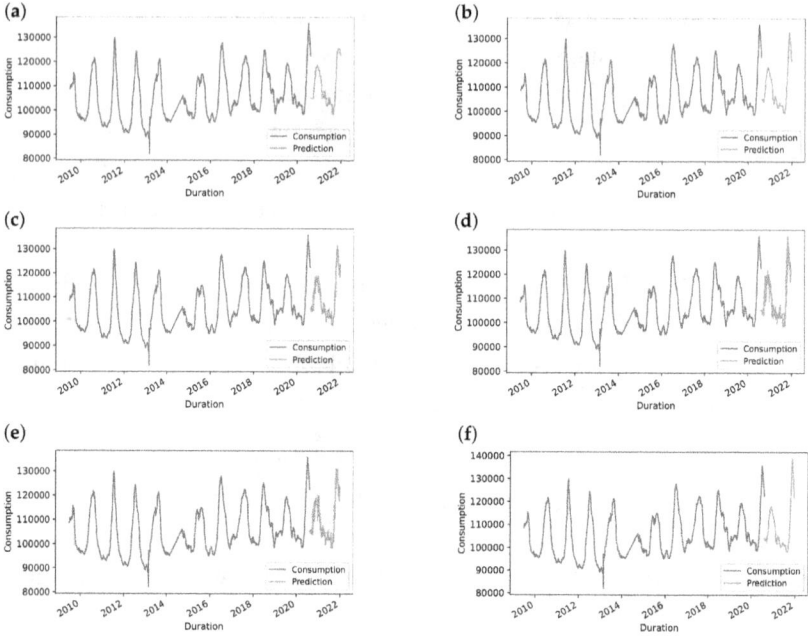

Figure 11.1   Water consumption predictions by Machine Learning models: (a) AdaBoost regressor; (b) linear regressor; (c) extra tree regressor; (d) decision tree regressor; (e) random forest regressor; and (f) support vector regressor (Rustam et al., 2022).

11.6 Machine Learning Models for Water Quality Prediction

ML algorithms, such as regression models, decision trees, and neural networks, are increasingly applied to predict water quality trends based on historical and current data. These models analyze patterns in a range of water quality parameters, including turbidity, nutrient levels, and pH, along with relevant environmental factors, such as rainfall, temperature, and land use.

For instance, ML models can predict the impact of agricultural runoff on water quality by analyzing how nutrient loading (e.g., nitrogen and phosphorus from fertilizers) interacts with rainfall patterns, soil conditions, and local topography. By learning from historical data, these models can identify predictive relationships that help forecast water quality conditions across seasons, regions, or even specific bodies of water.

For example, in river systems, ML models can evaluate how heavy rains or agricultural activities upstream influence turbidity or nutrient concentrations downstream. Predicting these changes allows authorities to anticipate and prepare for potential contamination events, such as algal blooms or eutrophication.

11.7 Real-Time Forecasting

An important advancement in predictive modeling is the integration of real-time data from IoT sensors into ML models for dynamic forecasting. Real-time forecasting enables continuous monitoring and provides immediate predictions regarding water quality trends, allowing for swift action to prevent contamination or environmental damage. For example, when continuous data streams from water quality sensors (monitoring parameters like temperature, pH, and DO) are combined with ML models, the resulting predictive systems can issue real-time forecasts. These forecasts may include the likelihood of algal blooms by predicting how current and future conditions—such as rising temperatures, increasing nutrient levels, and specific weather patterns—could foster the conditions for blooms. Such predictive capabilities are crucial for mitigating risks like oxygen depletion and toxin release from blooms, which can severely impact aquatic life and water safety.

11.8 Risk Assessment and Management

Predictive models are also essential for risk assessment and management, particularly in predicting contamination events and assessing potential impacts on water quality. By evaluating historical and current data trends, AI models can calculate the probability of various events leading to water contamination. This capability allows for a more proactive approach to water quality management, moving beyond reactive measures and toward preventing pollution before it happens. For example, ML models can forecast the impact of heavy rainfall on water quality by predicting how runoff will carry pollutants (e.g., sediments, chemicals, or waste) from land to water bodies. The model might assess various influencing factors, such as land use (urban or agricultural), topography, and weather conditions. When a model predicts high likelihoods of increased runoff, environmental managers can implement strategies such as adjusting water treatment processes, setting up temporary filtration systems, or enhancing watershed management to prevent contamination from reaching critical levels. These insights also enable agencies to plan for mitigation measures like water purification, emergency response protocols, or water conservation practices, especially in areas where contamination risk is elevated. For example, predictive models might forecast elevated levels of pollutants after a storm, allowing authorities to issue warnings or increase monitoring at key water sites.

11.9 Field Studies: Rivers, Lakes, and Groundwater Monitoring

Field studies play a crucial role in assessing the real-world effectiveness of water quality monitoring systems. These studies, particularly those utilizing IoT-based water quality monitoring, have provided critical insights into the dynamics of water pollution, contamination events, and the optimization of water management strategies. By deploying IoT sensors in natural water bodies, researchers and authorities can continuously monitor water quality parameters in rivers, lakes, and groundwater, offering real-time data that are essential for proactive management of water resources and pollution control.

11.10 Rivers and Lakes Monitoring

Many rivers and lakes face significant pollution challenges from agricultural runoff, industrial discharges, and the pressures of urbanization. In countries like India and China, major water bodies such as the Ganges River and Yangtze River have been equipped with IoT sensors to monitor water quality in real time. These sensors track critical water quality parameters such as turbidity, pH, DO, and heavy metals—each of which plays a crucial role in identifying pollution sources and understanding water body health.

For example, in the Ganges River, agricultural runoff, untreated sewage, and industrial effluents contribute to fluctuating pollution levels. IoT sensors deployed along various stretches of the river allow authorities to track pollution trends continuously. The data collected are fed into AI models that analyze the information and predict future water quality trends, helping to identify potential hazards before they worsen. These real-time data can trigger immediate responses, such as alerting local authorities to pollution spikes or guiding the optimization of water treatment processes to ensure the water is safe for human consumption and ecosystem health.

11.11 Groundwater Monitoring

Groundwater is a critical source of drinking water, particularly in rural and urban areas. However, it is vulnerable to contamination from sources like agricultural runoff, industrial waste, and wastewater disposal. IoT-based monitoring systems are increasingly used to detect contaminants in groundwater, providing a continuous assessment of water quality. These systems measure key parameters such as nitrate concentration, salinity, and arsenic levels, all of which are important indicators of contamination. For example, in agricultural regions where nitrate contamination is common due to fertilizer use, IoT sensors installed in wells can detect even small increases in nitrate levels, signaling potential contamination. This early detection enables authorities to implement mitigation measures such as limiting the use of fertilizers, improving irrigation practices, or enhancing waste management systems to reduce runoff. By providing data in real-time, IoT-based systems help prevent contamination from spreading and support long-term sustainability of groundwater resources.

11.12 Case Studies

11.12.1 The Amazon River Basin

In the Amazon River Basin, one of the most biodiverse and ecologically sensitive regions of the world, IoT-based water quality monitoring has been deployed across various tributaries to assess the impacts of deforestation, mining, and agriculture on water quality. In this case, sensors continuously measure water temperature, pH, turbidity, and DO levels, all of which are key indicators of water health.

AI-driven models analyze the data to detect pollution trends and forecast the long-term impact of human activities on water quality.

For example, deforestation and mining can introduce sediments and chemicals into the water, disrupting ecosystems and water quality. By identifying these risks early, policymakers can implement regulations on water management and pollution control to mitigate adverse impacts. These real-time data have been instrumental in shaping conservation policies aimed at protecting the Amazon's waterways and ensuring that local communities continue to have access to safe, clean water.

11.12.2 The Great Lakes (USA and Canada)

The Great Lakes, a vital freshwater resource for millions of people in both the USA and Canada, have been the subject of numerous field studies focused on IoT-based water quality monitoring. These lakes are under threat from pollution due to factors like agricultural runoff, the spread of invasive species, and the impacts of climate change. To combat these threats, IoT sensors continuously monitor key water quality parameters such as turbidity, nutrient concentrations, and the presence of harmful algal blooms (HABs), which pose significant environmental and public health risks. One of the significant advantages of IoT monitoring in the Great Lakes is its ability to forecast the occurrence of HABs. By analyzing real-time data related to water temperature, nutrient levels, and weather patterns, AI models can predict when and where harmful blooms are likely to form. This allows for early intervention and public advisories to protect human health and aquatic life. Additionally, IoT systems have helped identify sources of pollution and guided efforts to improve water quality through better agriculture management, pollution control, and habitat restoration.

12

SOIL HEALTH AND AGRICULTURAL APPLICATIONS

12.1 IoT-Enabled Soil Sensors for Precision Farming

Soil health is a critical determinant of agricultural productivity and environmental sustainability. Traditionally, soil monitoring was a labor-intensive process, involving periodic sampling and laboratory analysis. These methods often lacked the real-time capability required for effective decision-making in dynamic agricultural environments. However, the advent of Internet of Things (IoT)-enabled soil sensors has revolutionized soil health monitoring, providing continuous, real-time data that allow farmers and agricultural professionals to make timely and informed decisions.

12.2 Soil Sensors for Key Parameters

IoT-enabled soil sensors are designed to measure a variety of important soil characteristics, enabling farmers to gain a comprehensive understanding of soil health. Key parameters monitored by these sensors include:

- **Moisture Content**: Soil moisture is crucial for crop growth, as both over-watering and under-watering can negatively impact yields. IoT sensors help farmers track moisture levels at different soil depths, providing insights into the water availability in the soil. By monitoring moisture in real time, farmers can adjust irrigation practices to ensure optimal moisture levels, preventing water waste and reducing irrigation costs.
- **Temperature**: Soil temperature affects seed germination, root development, and nutrient uptake by plants. IoT sensors provide real-time data on soil temperature, helping farmers

DOI: 10.1201/9781003712701-15

understand the microclimate in their fields and make better decisions on planting schedules and crop selection.

- **pH Levels**: Soil pH influences nutrient availability and microbial activity in the soil. Monitoring pH levels with IoT sensors allows farmers to manage soil health more effectively, making adjustments such as adding lime or sulfur to adjust pH and optimize nutrient uptake for crops.
- **Salinity**: Excessive salinity can hinder plant growth and decrease agricultural productivity. IoT sensors monitor salinity levels, providing farmers with valuable information to take corrective actions such as leaching or soil amendments to reduce salt concentrations and improve soil quality.
- **Nutrient Concentrations**: IoT soil sensors measure the concentrations of key nutrients, such as nitrogen, phosphorus, and potassium, all of which are essential for plant growth. Real-time data on nutrient levels allow farmers to manage fertilization schedules and quantities, ensuring plants receive the nutrients they need for optimal growth.

These sensors are often placed at various soil depths to capture a detailed profile of soil conditions at different levels, providing a holistic view of the soil's suitability for crop growth. This continuous data stream is invaluable for assessing the overall health of the soil, which directly impacts crop productivity and sustainability.

12.3 Wireless and Remote Monitoring

One of the primary advantages of IoT-enabled soil sensors is their ability to wirelessly transmit data to centralized platforms. This connectivity allows farmers to monitor their fields remotely, reducing the need for frequent, manual inspections. The ability to access soil data from anywhere in real time enables farmers to act quickly and efficiently, responding to changing conditions as they arise. For example, if a sensor detects a sudden drop in soil moisture levels in a specific area, farmers can receive instant alerts and make adjustments to their irrigation system. This remote monitoring capability not only saves time and labor costs but also empowers farmers to make more informed decisions based on accurate, up-to-date data, leading to better resource management and improved productivity. The continuous flow of data also makes it easier for farmers to track long-term trends in soil health, helping them identify potential issues, such as nutrient depletion or soil degradation, before they become critical problems. By analyzing historical and real-time data, farmers can adjust their farming practices to maintain or improve soil health over time, promoting long-term sustainability.

12.4 Precision Irrigation and Fertilization

One of the most transformative applications of IoT-enabled soil sensors is in precision farming, which uses real-time soil data to optimize farming practices. Precision farming involves tailoring irrigation, fertilization, and other agricultural practices to the specific needs of the crops and the soil. This approach helps minimize resource waste, reduce environmental impact, and improve crop yields.

- **Precision Irrigation**: Soil moisture sensors are particularly useful in precision irrigation systems, where they detect areas of a field that are either over-irrigated or under-irrigated. By providing real-time insights into the moisture levels at various soil depths, IoT sensors allow farmers to apply water only where it is needed, reducing water usage and ensuring more efficient irrigation practices. This water conservation is crucial in areas facing water scarcity, and it helps farmers lower irrigation costs.
- **Precision Fertilization**: Fertilization is another area where IoT sensors play a critical role. Sensors monitor soil nutrient levels, allowing farmers to apply fertilizers only when necessary and in the correct amounts. This targeted fertilization reduces fertilizer waste, minimizes runoff, and prevents the environmental degradation caused by excessive chemical use. By matching fertilizer application to the specific needs of the soil and crops, farmers can improve crop yields while also reducing their environmental footprint.

Through precision farming, farmers can adopt a more data-driven approach to agriculture, where every decision is based on accurate, real-time data rather than general assumptions or outdated information. This results in optimized resource usage, cost savings, and enhanced crop productivity, contributing to the overall sustainability of farming practices.

12.5 AI Models for Yield Prediction and Soil Restoration

Artificial Intelligence (AI) has become an essential tool in modern agriculture, especially when integrated with data from IoT sensors. By processing and analyzing vast amounts of soil and environmental data, AI models can make informed predictions and recommendations that help farmers optimize their practices, improve productivity, and sustain soil health. These models play a crucial role in both yield prediction and soil restoration, two vital components of sustainable agriculture.

12.6 Yield Prediction Models

AI-powered yield prediction models use advanced algorithms to forecast crop productivity based on a variety of influencing factors. These models typically combine historical data (past yields, soil conditions) with real-time data (soil moisture, temperature, rainfall) to make predictions that can guide farming decisions. Common AI techniques used in yield prediction include:

- **Regression Analysis**: This statistical technique helps determine relationships between variables, such as soil health indicators and crop yields. By analyzing past data, regression models can estimate how specific soil conditions (e.g., pH, moisture) influence crop productivity.
- **Support Vector Machines (SVM)**: SVMs are effective for classifying and predicting outcomes based on complex datasets. In agriculture, they can analyze multidimensional data from soil sensors, weather patterns, and historical crop yields to predict how changes in environmental conditions will affect the crop.
- **Neural Networks**: These AI models simulate the way the human brain works by learning from large datasets. Neural networks can

handle more complex and non-linear relationships between factors like soil composition, crop types, weather patterns, and agricultural practices. This makes them highly effective at predicting how different variables interact to impact crop yields.

By using these AI models, farmers can predict not just average yields but also the potential outcomes of various farming practices. For instance, AI can forecast how different soil management techniques, such as adjusting pH or improving soil structure, will affect future crop yields. These predictions allow farmers to make data-driven decisions about which crops to plant, when to plant them, and how to allocate resources like water and fertilizers.

12.7 Soil Restoration through AI

AI also plays a vital role in soil restoration by analyzing long-term soil health data and providing insights into the causes and consequences of soil degradation. Through the use of AI-driven models, farmers can identify trends such as nutrient depletion, erosion, or contamination that lead to soil deterioration.

- **Detecting Degradation**: AI can process data collected over time, such as changes in soil pH, organic matter content, and nutrient levels. By identifying patterns in these data, AI can highlight areas where soil health is deteriorating. This early detection helps farmers take preventative or corrective actions before the soil reaches critical levels of degradation.
- **Restoration Recommendations**: AI models can recommend effective soil management techniques based on the specific degradation issues identified. For example, if nutrient depletion is detected, AI may suggest practices such as crop rotation to restore nitrogen levels or the addition of organic matter to improve soil structure and fertility. Similarly, AI models can recommend reduced tillage to prevent further soil erosion or introduce cover crops to enhance soil stability.
- **Long-Term Impact Predictions**: One of the strengths of AI is its ability to simulate the long-term impacts of different interventions. By analyzing how past soil restoration efforts

have affected soil health, AI models can predict the effectiveness of proposed restoration strategies. This helps farmers choose the most sustainable practices that will improve soil quality over time, leading to sustainable agriculture.

12.8 Optimizing Fertilizer and Irrigation Use

AI models are particularly effective at optimizing the use of fertilizers and irrigation, both of which are critical components of modern agriculture.

- **Fertilizer Optimization**: AI analyzes soil nutrient data and recommends the precise amount of fertilizer needed to achieve optimal crop growth. Over-fertilization can lead to waste and environmental pollution, while under-fertilization can limit crop yields. AI models help find the optimal balance, reducing costs and improving environmental sustainability. For example, an AI system could recommend the best time to apply specific fertilizers based on current soil nutrient levels and weather conditions.
- **Irrigation Scheduling**: Efficient irrigation is vital for maximizing crop productivity while conserving water. AI can integrate real-time soil moisture data from sensors with weather forecasts to generate precise irrigation schedules. This ensures that crops receive the right amount of water at the right time, avoiding over-irrigation (which can lead to waterlogging and nutrient leaching) or under-irrigation (which can cause drought stress). AI models can even adapt irrigation schedules based on changing weather conditions, further optimizing water usage and promoting sustainability.

By integrating AI into fertilizer and irrigation management, farmers can reduce their environmental footprint, lower costs, and boost crop productivity. These AI-driven approaches ensure that resources are used efficiently, contributing to a more sustainable agricultural system.

12.9 Practical Case Studies in Sustainable Agriculture

Several real-world case studies showcase how the integration of IoT and AI technologies is revolutionizing sustainable agriculture. These examples illustrate how these innovative technologies are enhancing soil health management, optimizing crop productivity, and fostering environmentally friendly farming practices across various regions. Each case highlights the practical applications of IoT and AI to address the unique challenges faced by farmers worldwide, while promoting sustainability in agriculture.

12.10 Case Studies

12.10.1 Precision Agriculture in the Netherlands

The Netherlands, known for its highly efficient and technologically advanced agricultural systems, has been a leader in the adoption of precision farming techniques. In this case study, IoT soil moisture sensors are deployed across expansive agricultural fields to monitor the moisture levels of the soil in real-time. These sensors continuously measure soil moisture content and transmit the data to a centralized system, where the data are processed by AI models. The AI models analyze these data to predict the specific water requirements of different crops at any given time. This predictive capability enables farmers to apply precise irrigation, reducing water waste and ensuring that crops receive the exact amount of moisture needed for optimal growth. The use of IoT and AI in this context also extends to fertilizer management. AI models evaluate real-time data from soil sensors regarding nutrient levels (such as nitrogen, phosphorus, and potassium) and recommend the most efficient application of fertilizers. This approach not only enhances crop yields but also minimizes the environmental impact by reducing the excess use of fertilizers, which can otherwise lead to soil and water pollution. Overall, this precision farming approach in the Netherlands is a model for sustainable agricultural practices that conserve resources and protect the environment.

12.10.2 Smart Agriculture in India

In rural India, where agriculture is a significant part of the economy and water scarcity remains a pressing issue, IoT-enabled soil sensors and AI models are transforming farming practices. In regions where water availability is limited, soil moisture sensors are used to continuously monitor the soil's moisture levels. Based on the data collected, AI models predict irrigation needs, allowing farmers to optimize water usage by ensuring that water is applied only when and where it is needed. This helps mitigate water wastage and ensures the health of crops, especially in dry regions. Moreover, AI-driven yield prediction models in India analyze real-time data from the soil sensors and environmental factors such as temperature and rainfall. These data enables farmers to predict crop yields more accurately and plan their harvests efficiently. Additionally, AI models provide actionable insights into soil restoration techniques. For example, recommendations may include the use of organic fertilizers and the practice of contour farming to reduce soil erosion. By utilizing these technologies, farmers can not only enhance productivity but also restore soil health over time, promoting sustainable agricultural practices that align with environmental goals.

12.10.3 Soil Health Monitoring in the United States

In the United States, especially in states like California where agriculture is vital to the economy, IoT and AI are being leveraged to improve soil health and optimize farming practices. In one example, a farming company in California uses an extensive network of IoT soil sensors to monitor moisture, temperature, and salinity in real-time across multiple fields. These data are fed into AI models that predict the optimal irrigation schedules and fertilization needs based on the real-time soil conditions. By aligning water and fertilizer application with crop requirements, the system reduces both water usage and fertilizer costs, leading to cost savings and environmental benefits. Furthermore, the system contributes to soil health restoration by suggesting best practices for crop rotation and reduced tillage. These strategies help maintain the fertility and structure of the soil over the long term, preventing degradation and supporting sustainable farming.

As a result, the farming company not only improves productivity but also promotes a sustainable approach to agriculture that protects the land for future generations.

12.10.4 Sustainable Farming in Kenya

In Kenya, IoT and AI technologies are being applied to enhance soil health and agricultural productivity, particularly on smallholder farms. These technologies are designed to help farmers who often face challenges such as poor soil fertility, inadequate water resources, and limited access to modern farming techniques. Through the use of soil sensors, farmers are able to monitor key soil parameters like moisture content and nutrient levels in real-time. AI models analyze these data to predict potential crop yields and provide recommendations for improving soil health. For instance, AI might suggest the use of organic fertilizers to reduce dependency on chemical inputs, or offer advice on soil restoration techniques such as increasing organic matter or practicing agroforestry. This integrated approach helps farmers to reduce chemical fertilizer and pesticide use, promoting organic farming practices that improve the long-term fertility and sustainability of the soil. By optimizing resource usage and making data-driven decisions, Kenyan farmers can increase productivity while minimizing their environmental footprint. This also enhances profitability, which contributes to the economic stability of smallholder farms, thereby improving food security and supporting local communities.

13

DISASTER MONITORING AND CLIMATE ANALYSIS

13.1 IoT Solutions for Early Warning Systems (Floods, Fires, etc.)

Natural disasters such as floods, fires, droughts, and hurricanes pose significant threats to human life, property, and the environment. Early warning systems play a critical role in mitigating disaster impacts by providing timely alerts, allowing people and authorities to take preventive actions. Internet of Things (IoT)-based solutions have become key components of modern disaster monitoring systems, enabling real-time data collection, processing, and communication.

- **Flood Monitoring Systems**: IoT-based flood monitoring systems deploy sensors at critical locations such as riverbanks, dams, and urban areas prone to flooding. These sensors monitor water levels, rainfall, soil moisture, and river flow rates. For example, IoT-enabled sensors placed along river systems can detect rapid changes in water levels, which may indicate the potential for flooding. The data are transmitted in real-time to centralized platforms for analysis, and Artificial Intelligence (AI) models are used to predict flood events, issue warnings, and optimize flood control measures.
- **Wildfire Detection Systems**: Wildfires, which are increasingly common due to climate change, can spread rapidly, making early detection crucial for saving lives and preventing large-scale damage. IoT sensors, such as temperature, humidity, and smoke detectors, can be deployed in forests or other vulnerable areas to monitor fire risks. These sensors can detect even small changes in environmental conditions that may signal the onset of a wildfire. Combined with AI models that analyze sensor data, authorities can issue early warnings and activate firefighting resources more efficiently.

DOI: 10.1201/9781003712701-16

- **Other Disasters (Tornadoes, Landslides, Tsunamis)**: IoT-based monitoring systems are also used for other natural disasters, such as tornadoes, landslides, and tsunamis. For instance, ground vibration sensors and soil moisture sensors can help detect signs of potential landslides, while ocean-based sensors can monitor seismic activity and rising ocean levels to provide early warning for tsunamis. Integrating these IoT systems into a broader disaster management framework enables quicker responses and better preparedness.

13.2 AI Applications in Climate Pattern Recognition

AI has the potential to significantly enhance the accuracy and efficiency of climate pattern recognition. By analyzing large amounts of climate data, AI models can identify trends, predict future climate events, and provide valuable insights for disaster preparedness and mitigation (Table 13.1).

Table 13.1   AI Applications in Climate Pattern Recognition

APPLICATION	AI METHODS USED	SPECIFIC EXAMPLES
Climate modeling and prediction	• Machine learning (ML) • Neural networks • Reinforcement learning	• *DeepMind's Earth Engine*: Simulates temperature changes and climate patterns. • Predicting monsoon and drought cycles.
Extreme weather event detection	• Deep learning • Convolutional neural networks (CNNs) • Time series ML	• AI predicts hurricanes by analyzing atmospheric pressure and wind speed data. • Heatwave risk forecasting.
Remote sensing and satellite data	• Computer vision • Random Forests • Generative adversarial networks (GANs)	• AI analyzes satellite images to detect deforestation • Tracks sea-ice melting using remote sensing.
Carbon emission tracking	• Natural language processing (NLP) • Deep reinforcement learning (DRL)	• AI-powered satellites detect global CO_2 emission hotspots. • Tracking methane emissions using drones and Machine Learning algorithms.

(Continued)

Table 13.1 (*Continued*)   AI Applications in Climate Pattern Recognition

APPLICATION	AI METHODS USED	SPECIFIC EXAMPLES
Ocean and atmospheric monitoring	• Recurrent neural networks (RNNs) • Physics-informed neural networks (PINNs)	• Monitoring El Niño/La Niña phenomena. • Predicting changes in ocean currents (e.g., Gulf Stream analysis).
Wildfire prediction and monitoring	• Decision trees • Gradient boosting algorithms • AI-powered drones	• *FireCast*: Predicts wildfire risks using satellite imagery and vegetation data. • Forecasting wildfire spread based on wind patterns.
Glacier and ice monitoring	• Image segmentation • 3D point cloud analysis	• AI models track glacier retreat using satellite images. • Analyzing Arctic Sea ice thickness over time.
Biodiversity and ecosystem analysis	• Support vector machines (SVMs) • Clustering algorithms	• AI identifies shifts in biodiversity patterns due to climate change. • Tracks species migration paths.

- **AI in Climate Modeling**: AI algorithms, such as deep learning and neural networks, are used to process vast datasets from various sources, including satellites, weather stations, and ocean buoys. These models analyze climate variables such as temperature, wind patterns, atmospheric pressure, and precipitation to recognize trends and predict future climate conditions. For example, AI models can predict the likelihood of extreme weather events, such as heatwaves, storms, and droughts, based on historical data and current climate conditions.

- **Predicting Long-Term Climate Changes**: AI is also used for long-term climate pattern recognition, such as forecasting changes in global temperatures, sea level rise, and shifts in rainfall patterns due to climate change. These predictive models can help policymakers and scientists understand potential impacts on agriculture, water resources, and coastal regions. Machine Learning models can incorporate complex variables, including greenhouse gas emissions, land use changes, and energy consumption, to provide more accurate climate predictions.

- **Real-Time Climate Monitoring**: Real-time climate monitoring systems use AI to analyze sensor data from IoT devices, satellites, and other climate monitoring tools. These AI-powered systems can detect anomalies in climate patterns, such as sudden temperature spikes, unusual precipitation patterns, or shifts in wind direction, that may indicate the onset of extreme weather events. By providing early warning capabilities, AI models help improve response times and enhance disaster preparedness.

13.3 Examples: Monitoring Droughts, Hurricanes, and Heatwaves

AI and IoT technologies have been applied to monitor and manage extreme weather events, such as droughts, hurricanes, and heatwaves, which have become more frequent and intense due to climate change. Below are some examples of how these technologies are being used in practice.

- **Drought Monitoring**: Droughts are prolonged periods of insufficient rainfall that can have devastating effects on agriculture, water supplies, and ecosystems. IoT sensors, including soil moisture sensors, weather stations, and remote sensing devices, can monitor drought conditions by providing real-time data on soil moisture levels, rainfall, and temperature. AI models analyze these data to predict drought events and assess their potential severity. For example, in parts of Africa and Asia, IoT-based drought monitoring systems help inform water management decisions, optimize irrigation systems, and prepare for water shortages. AI-powered models predict when droughts are likely to occur and assess their impact on crop production, water resources, and human populations.

- **Hurricane Monitoring**: Hurricanes are among the most destructive natural disasters, and accurate predictions are essential for reducing damage and saving lives. IoT systems are used to collect data on ocean temperatures, wind speed, atmospheric pressure, and rainfall from sensors placed on ships, buoys, and satellites. These data feed into AI models that analyze storm patterns, track hurricane development, and predict landfall locations. AI models also estimate the potential strength of hurricanes, allowing emergency management teams to prepare accordingly. For instance, the National Hurricane Centre in the U.S. uses AI algorithms to improve the accuracy of hurricane forecasting, providing critical information for evacuation orders and resource allocation.

- **Heatwaves**: Heatwaves, characterized by prolonged periods of excessively hot weather, are a growing concern due to global warming. IoT sensors, including temperature sensors and humidity detectors, monitor real-time weather conditions and provide early warnings of heatwaves. AI models analyze historical and real-time climate data to predict heatwave occurrences and their potential impacts on human health, agriculture, and infrastructure. In cities, IoT systems can be used to monitor urban heat islands—areas with significantly higher temperatures than their surroundings—allowing for more targeted cooling interventions, such as increasing green spaces or installing reflective materials. AI-driven heatwave prediction systems are already in use in cities like Paris, which has implemented a heat action plan based on AI forecasts to protect vulnerable populations.

By combining IoT and AI, disaster monitoring and climate analysis systems offer unprecedented opportunities to enhance preparedness, response, and resilience to extreme weather events. These technologies provide real-time insights, predictive capabilities, and early warnings that enable authorities and communities to take proactive measures, saving lives, minimizing damage, and supporting sustainable development in the face of increasing climate risks.

13.4 Case Studies

1. **Amazon Rainforest—Colombia Joint Platform**

   Researchers from the University of the Andes, Humboldt Institute, and SINCHI, in partnership with Microsoft, use AI to analyze satellite images, bioacoustic recordings, and camera trap data. Their platform monitors deforestation and biodiversity, helping protect this vital ecosystem.

2. **South Africa—Deep Learning for Savanna Vegetation**

   With backing from Microsoft's AI for Earth program, the GeoAI Lab creates spatial prediction tools to assess plant vitality in the Cape Floristic Region, an area that contains nearly a fifth of Africa's biodiversity.

3. **SPARROW—Biodiversity in Remote Areas**

   SPARROW, an autonomous solar-powered device created through Microsoft AI for Good, uses sound and image sensors to track wildlife. Initial trials in Ecuador and Colombia provide satellite-connected, real-time monitoring of biodiversity.

4. **Environmental Pollution Monitoring**

   - **CleanHub** leverages AI to evaluate geo-located images of plastic litter, offering transparent and immediate updates on waste cleanup efforts in developing regions.
   - In West Africa, UNESCO equips journalists with AI tools to identify and report on industrial pollution across Ghana, Cameroon, Gabon, and Nigeria.

5. **Natural Disaster Prevention**

   - **SERVIR Mekong**, a joint effort by USAID and NASA, applies AI alongside satellite data to monitor flooding, land usage, and environmental conditions across Cambodia, Laos, Myanmar, Thailand, and Vietnam.
   - Microsoft AI for Good labs based in Nairobi and Cairo assist in Africa by providing early warnings for droughts and enhancing water resource management in the region.

6. **Carbon Capture and Climate**

   - **Albo Climate**, an Israeli startup operating in Africa and Asia, focuses on tracking land use and measuring carbon emissions using AI-driven satellite imagery.
   - **Ant Forest** in China promotes eco-friendly habits, engaging 650 million participants to plant trees and reduce carbon dioxide emissions by 26 million tons.

13.5 Sample of Data Sets

Below are examples of commonly used datasets in AI-driven environmental research, particularly for biodiversity, pollution tracking, climate
analysis, and remote sensing. Many are openly accessible and valuable
for scientists globally, including those from underrepresented areas:

13.5.1 Biodiversity and Wildlife Monitoring

1. **Snapshot Serengeti**
   - **Site**: Serengeti National Park, Tanzania
   - **Dataset**: Millions of camera trap photos labeled by species
   - **Purpose**: Developing AI models to identify species and
     study their behavior
   - **Availability**: snapshotserengeti.org/data
2. **iNaturalist Dataset**
   - **Data**: Worldwide images of plants, animals, and fungi
     contributed by citizen scientists
   - **Application**: Identifying species and mapping their geo-
     graphic distribution
   - **Access**: inaturalist.org/datasets
3. **BirdCLEF Datasets**
   - **Dataset**: Bird call audio clips collected from various places
     worldwide
   - **Purpose**: Using AI to identify species through sound
     analysis
   - **Access**: birdclef.org

13.5.2 Satellite and Remote Sensing Data

1. **Sentinel-2 Satellite Imagery**
   - **Data**: Multispectral satellite imagery capturing land use,
     vegetation, and water features
   - **Application**: Classifying land cover and tracking deforestation
   - **Access**: esa.int/Applications/Observing_the_Earth/
     Copernicus/Sentinel-2

2. **Landsat Data Archive**
   - **Dataset**: Earth surface images spanning from 1972 to present
   - **Purpose**: Studying climate change effects and urban growth
   - **Access**: landsat.gsfc.nasa.gov
3. **Global Forest Watch**
   - **Data**: Worldwide alerts on deforestation and changes in forest coverage
   - **Application**: Monitoring forest loss and assessing ecosystem conditions
   - **Access**: globalforestwatch.org

13.5.3 Pollution and Waste Tracking

1. **PlasticNet Dataset**
   - **Dataset**: Photos and related data of plastic waste found in marine and coastal areas
   - **Purpose**: Teaching AI systems to detect plastic pollution in imagery
   - **Access**: plasticnet.org
2. **Air Quality Open Datasets**
   - **Examples**: OpenAQ and the U.S. EPA's Air Quality System
   - **Application**: Modeling and predicting air pollution levels
   - **Access**: openaq.org

13.5.4 Climate and Weather Data

1. **NOAA Climate Data**
   - **Data**: Global records of weather conditions, temperature, and rainfall
   - **Purpose**: Analyzing climate trends and making forecasts
   - **Access**: noaa.gov/data
2. **ERA5 Reanalysis Dataset**
   - **Dataset**: Detailed global atmospheric measurements with fine spatial and time scales
   - **Application**: Studying climate dynamics and forecasting extreme weather events
   - **Access**: cds.climate.copernicus.eu

PART IV

Broader Perspectives and Challenges

14

ETHICS AND SOCIAL IMPLICATIONS

14.1 Ethical Issues in AI Decision-Making

As Artificial Intelligence (AI) becomes more integrated into environmental monitoring and management systems, it brings up important ethical questions related to the role of AI in decision-making. AI models are often used to analyze vast amounts of environmental data, make predictions, and suggest interventions. However, these models are only as good as the data they are trained on, and the decision-making process can sometimes be opaque, leading to concerns about accountability, transparency, and fairness.

- **Accountability and Transparency**: One of the key ethical concerns in AI decision-making is accountability. In environmental monitoring, AI algorithms may determine policies for managing natural resources, predicting disaster risks, or assessing environmental health. If these decisions are wrong or lead to harmful outcomes, who is accountable? The opacity of some AI systems, especially deep learning models, makes it difficult for decision-makers to understand how conclusions are drawn. Ensuring transparency in the design and operation of AI models, and allowing human oversight, is essential for preventing unintended consequences and ensuring that AI systems are used responsibly.
- **Autonomy vs. Human Control**: AI has the potential to automate many aspects of environmental monitoring, including disaster response, resource management, and pollution control. However, this raises concerns about the loss of human control over critical decisions, particularly in situations where rapid responses are needed, such as during natural disasters. While AI can assist in decision-making, ensuring that human

DOI: 10.1201/9781003712701-18

expertise remains integral to the process is vital to ensure ethical governance and appropriate interventions.

- **Ethical Considerations in Decision Algorithms**: The algorithms used in AI decision-making often reflect the values and assumptions of the individuals or organizations who design them. Ethical considerations, such as equity, environmental justice, and sustainability, must be integrated into AI models to avoid reinforcing negative biases or prioritizing certain groups over others. For example, AI algorithms used for climate change mitigation strategies should consider the social and economic impacts on vulnerable populations, ensuring that climate actions do not disproportionately harm marginalized communities.

14.2 Data Privacy and Security in IoT Systems

Internet of Things (IoT)-based environmental monitoring systems rely heavily on collecting and transmitting large amounts of real-time data from sensors placed in various locations. While these data are essential for tracking environmental conditions and making informed decisions, it raises significant concerns about data privacy, security, and ownership.

- **Data Privacy**: Environmental IoT systems often collect sensitive data, such as information about air quality, water resources, and agricultural practices, which may include location data, personal health information, or proprietary business data. Protecting these data from unauthorized access or misuse is a critical challenge. Additionally, as more smart devices are integrated into the environment, there is a growing risk that individuals' personal data may be collected inadvertently or without their consent. Governments and organizations must establish robust data privacy regulations and frameworks to protect citizens' rights and ensure responsible data collection and usage. Figure 14.1 describes the various steps of data privacy and security in IoT systems.
- **Cybersecurity Risks**: With the proliferation of IoT devices, the risk of cyberattacks targeting environmental monitoring systems has increased. These systems are often connected

Ensuring Data Privacy in IoT Systems

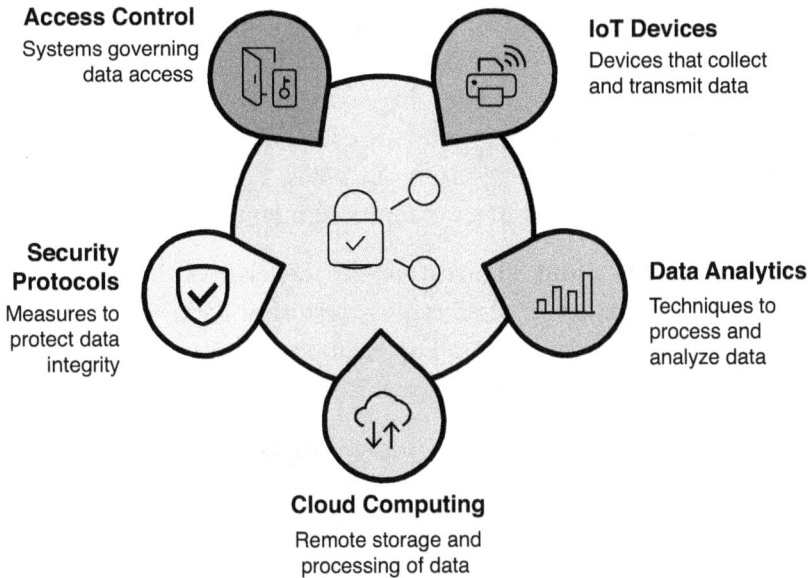

Access Control
Systems governing
data access

IoT Devices
Devices that collect
and transmit data

**Security
Protocols**
Measures to
protect data
integrity

Data Analytics
Techniques to
process and
analyze data

Cloud Computing
Remote storage and
processing of data

Figure 14.1   Data privacy and security in IoT systems. (Created by the authors.)

to larger networks, which may be vulnerable to hacking or data breaches. In the case of disaster monitoring systems, for example, a cyberattack could disrupt the ability to issue timely alerts or interfere with emergency response efforts. Securing IoT systems and ensuring their resilience to cyber threats is essential for maintaining the integrity of environmental monitoring efforts.

- **Data Ownership and Access**: Another issue related to data privacy and security is the question of data ownership. In many IoT-based environmental systems, data are generated by a variety of stakeholders, including individuals, governments, businesses, and research institutions. Determining who owns the data, who has access to it, and how it can be used for decision-making is a complex issue. Establishing clear protocols for data sharing, consent, and ownership is necessary to avoid conflicts and ensure equitable access to valuable environmental data.

14.3 Addressing Bias and Inclusivity in Environmental Monitoring

Environmental monitoring systems powered by AI and IoT have the potential to improve decision-making and promote sustainability. However, these systems can also perpetuate biases and inequalities if not designed inclusively. Addressing issues of bias and ensuring that environmental monitoring systems are equitable and inclusive are critical steps for achieving sustainable outcomes. Figure 14.2 shows how bias and inclusivity work in environmental monitoring.

- **Bias in Data and Algorithms**: AI models used in environmental monitoring often rely on historical data, which may contain inherent biases. For example, environmental data collected from specific geographic regions or socio-economic

Figure 14.2 Addressing bias and inclusivity in environmental monitoring. (Created by the authors.)

groups may not be representative of broader populations or ecosystems. If AI models are trained on biased data, they may produce inaccurate predictions or recommendations that disproportionately affect marginalized communities or vulnerable environments. Ensuring diversity in data collection, validation of AI models, and the use of fairness metrics are essential for mitigating bias.

- **Inclusivity in Decision-Making**: To promote equitable environmental monitoring, it is important to involve diverse groups of stakeholders in the design and implementation of IoT and AI systems. This includes ensuring that local communities, indigenous groups, and other marginalized populations have a voice in how data are collected, analyzed, and used for decision-making. Involving a wide range of stakeholders helps ensure that environmental policies are not only effective but also fair and inclusive.

- **Access to Technology**: One of the challenges in addressing bias and inclusivity in environmental monitoring is ensuring that the technology is accessible to all communities. Many rural or low-income areas may lack the infrastructure or resources to deploy IoT sensors or benefit from AI-powered environmental management systems. Bridging the digital divide by making these technologies more affordable, accessible, and adaptable to local contexts is crucial for promoting environmental justice.

- **Ethical Use of Environmental Data**: As environmental data are increasingly used to inform policy and business decisions, ethical concerns about the equitable distribution of resources and opportunities must be addressed. For instance, using AI models to monitor environmental impacts in developing countries must account for the capacity of local governments and communities to respond to these findings. Data-driven decisions that disproportionately affect low-income or vulnerable populations should be avoided. Ethical frameworks for data use and environmental policy are necessary to ensure that technology serves the public good.

The integration of AI and IoT technologies in environmental monitoring presents significant opportunities for improving sustainability,
disaster response, and resource management. However, these technologies also raise important ethical and social challenges that must
be carefully addressed. Ethical AI decision-making, data privacy and
security, and inclusivity in environmental monitoring are all essential
considerations for ensuring that these technologies are used responsibly
and equitably. By addressing these challenges, we can maximize the
potential of AI and IoT to promote a sustainable and just future for all.

15

EMERGING TRENDS AND FUTURE TECHNOLOGIES

15.1 Blockchain in Environmental Data Integrity

Blockchain technology, initially known for its application in crypto-currencies, is increasingly being explored for its potential in ensuring the integrity and transparency of environmental data. In environmental monitoring, data collected from various sources such as Internet of Things (IoT) sensors, satellite systems, and other monitoring tools must be reliable and tamper-proof to support critical decision-making processes. Blockchain provides a decentralized and immutable ledger, which can guarantee the authenticity and traceability of environmental data. Pros and cons of blockchain in environmental data integrity have been given in Table 15.1.

- **Ensuring Data Integrity**: One of the major concerns in environmental monitoring is ensuring that data remain unaltered or tampered with after collection. Blockchain's decentralized nature ensures that data once recorded cannot be altered or deleted without leaving a trace. This makes it ideal for applications such as monitoring air quality, water contamination, and deforestation, where data integrity is crucial for regulatory compliance and environmental protection.
- **Transparency and Trust**: Blockchain enables transparency by providing a shared ledger that allows all stakeholders, from governments to citizens, to access and verify environmental data. This can increase public trust in the data and their use, fostering better collaboration and accountability in environmental decision-making. For example, blockchain can be used to track carbon credits or emissions reductions, ensuring that claims made by companies or governments are verified and cannot be manipulated.

DOI: 10.1201/9781003712701-19

Table 15.1 Pros and Cons of Blockchain in Environmental Data Integrity

ASPECT	PROS	CONS
Transparency	Blockchain's immutable ledger ensures data cannot be tampered with, building trust among stakeholders. For example, organizations can track emissions data in real-time, reducing fraud in carbon credit trading systems (Copernicus Climate Change Service, 2023).	Transparency may expose sensitive environmental data, raising privacy or security concerns for vulnerable communities or politically sensitive regions.
Decentralization	No single authority controls the system, reducing risks of data manipulation or censorship. For instance, it ensures accurate reporting of GHG emissions without reliance on governments or corporations (Ham et al., 2019).	Effective implementation requires coordination among stakeholders, and disputes over governance can delay adoption (Ham et al., 2019; Liu et al., 2022).
Immutability	Once recorded, data cannot be altered, making it ideal for long-term storage (e.g., tracking sea-level rise over decades) and ensuring historical data integrity (Rasp et al., 2020).	Faulty or inaccurate data, such as from malfunctioning IoT sensors, become permanently stored, propagating inaccuracies (Rasp et al., 2020; IEA, 2023).
Auditability	Built-in audit trails make it easy to track data origins and ensure accountability, e.g., in carbon offset programs or renewable energy certifications (Urban et al., 2020).	Storing large datasets like satellite imagery is expensive, making on-chain solutions less practical for high-volume environmental data (Urban et al., 2020; NASA Cryosphere Program, 2023).
Energy efficiency	Emerging blockchain protocols like Proof of Stake (PoS) consume far less energy compared with traditional Proof of Work (PoW) systems, making them environmentally friendly.	Traditional PoW systems (e.g., Bitcoin) consume vast amounts of energy, undermining environmental goals. Bitcoin mining alone consumes more electricity annually than some small countries.
Scalability	Blockchain is suitable for smaller datasets, such as tracking sustainable product certifications or monitoring small-scale environmental projects.	Large-scale datasets, such as global climate models or IoT data streams, can overwhelm the blockchain, leading to inefficiency and high costs (Liu et al., 2022).
Integration with IoT	IoT sensors can feed real-time environmental data (e.g., water pollution levels) directly into blockchain, ensuring secure and reliable storage, and automating alerts for policy responses (Rasp et al., 2020).	IoT devices are susceptible to errors, tampering, or inconsistent readings, which, once uploaded, become immutable and could misinform decisions (Rasp et al., 2020; Copernicus Climate Change Service, 2023).

(Continued)

Table 15.1 (*Continued*)   Pros and Cons of Blockchain in Environmental Data Integrity

ASPECT	PROS	CONS
Cost	Blockchain automates processes like environmental monitoring, reducing administrative overheads and improving efficiency (Urban et al., 2020; Ham et al., 2019).	High initial implementation costs, including infrastructure and training, may limit adoption, especially in developing countries (Ham et al., 2019).
Standardization	Open-source blockchain fosters global standardization across regions, enabling interoperability, such as uniform reporting standards for deforestation monitoring.	A lack of established standards complicates cross-border environmental initiatives and hampers data sharing between systems (Liu et al., 2022).
Regulation	Transparent systems encourage better regulatory compliance, e.g., tracking companies' emissions and ensuring adherence to environmental laws (IEA, 2023).	Unclear or inconsistent regulations surrounding blockchain adoption (e.g., privacy laws like General Data Protection Regulation can delay its use in environmental monitoring projects (NASA Cryosphere Program, 2023).

- **Supply Chain Transparency**: Blockchain can also improve the sustainability of supply chains by providing real-time, verifiable information on the environmental impact of goods and services. For example, blockchain could track the carbon footprint of products from production to delivery, ensuring transparency and enabling consumers to make more informed, sustainable choices.
- **Integration of Autonomous Systems (Drones, Robots)**

Autonomous systems, including drones and robots, are becoming increasingly important in environmental monitoring and management. These systems enable the collection of real-time, high-resolution data in areas that may be difficult or dangerous for humans to access. They also offer the ability to conduct continuous, large-scale monitoring with minimal human intervention, providing more accurate and efficient data collection.

- **Drones in Environmental Monitoring**: Drones, or unmanned aerial vehicles, are already being used in a wide range of environmental applications, such as monitoring deforestation, mapping land degradation, and assessing air quality. Drones equipped with advanced sensors, cameras, and remote sensing technology can capture high-resolution images and data from remote or hard-to-reach areas. For instance, drones are used in environmental conservation to monitor wildlife habitats and track changes in vegetation cover, providing critical data for biodiversity protection.

- **Robots for Environmental Cleanup**: Autonomous robots, both ground-based and underwater, are being developed for environmental cleanup efforts, particularly in areas affected by pollution. For example, robots are being used to remove plastic waste from oceans or clear hazardous materials from contaminated soil. These robots can operate in hazardous environments, reducing the risk to human workers while improving efficiency and precision in environmental remediation tasks.

- **Collaboration with Artificial Intelligence (AI)**: The integration of AI with autonomous systems allows drones and robots to process data in real time, make decisions based on pre-defined algorithms, and adapt to changing conditions. AI-enabled autonomous systems can detect patterns in environmental data and make decisions without human intervention, such as identifying sources of pollution or determining the best locations for deploying sensors. This combination of AI, robotics, and IoT will significantly enhance the capabilities of environmental monitoring systems in the future.

15.2 Advances in IoT with Edge and Cloud Computing

As IoT systems continue to evolve, the integration of edge and cloud computing is poised to revolutionize environmental monitoring by improving the efficiency, scalability, and real-time processing capabilities of these systems.

- **Edge Computing**: Edge computing refers to processing data locally on IoT devices, rather than transmitting all data to a central server or cloud for processing. This approach reduces latency, conserves bandwidth, and enables faster decision-making in real-time. In environmental monitoring, edge computing can be particularly valuable for applications that require immediate responses, such as flood detection, air quality management, and wildlife monitoring. For example, sensors deployed in a forest to detect wildfires can analyze temperature and smoke data at the edge, triggering an immediate alarm if fire conditions are detected, without waiting for data to be sent to the cloud. Table 15.2 shows the challenges and future research directions for edge computing.

Table 15.2 Summary of Challenges and Future Research Directions for Edge Computing (EC)-IoT Security

CHALLENGE	DESCRIPTION	FUTURE RESEARCH DIRECTIONS
Power constraints	Limited processing power in edge devices requires solutions like model pruning, on-device learning, and federated learning (FL), which can reduce data transmission but require complex implementation.	• Develop advanced pruning algorithms to maintain accuracy • Enhance model efficiency with quantization and knowledge distillation • Design energy-efficient FL algorithms • Integrate edge caching and opportunistic computing
Training constraints	Task offloading helps with intensive tasks but depends on network reliability and may introduce latency. Edge-centric training can enhance autonomy and efficiency.	• Create efficient task offloading strategies • Develop edge-centric training techniques to reduce cloud dependency
Memory limitations	Quantization reduces memory usage but can affect model precision. Mixed precision training can optimize performance and memory efficiency.	• Enhance post-training quantization methods • Develop mixed precision training techniques
Local processing risks	Hybrid processing ensures real-time processing and reduces latency but requires sophisticated architecture.	• Develop optimized hybrid architectures • Implement dynamic task allocation algorithms
Data privacy	Edge–cloud collaboration enhances data safety but can introduce latency and requires robust communication channels.	• Enhance secure multi-party computation methods • Leverage blockchain technology for secure data transactions
Ethical concerns	AI integration in EC-IoT raises ethical concerns such as potential bias, transparency, accountability, and informed consent.	• Address AI bias and ensure fairness • Ensure transparency and accountability in AI decision-making • Ensure informed consent and clear communication • Develop ethical guidelines prioritizing user well-being

- **Cloud Computing**: Cloud computing, on the other hand, allows for the centralized storage and processing of vast amounts of environmental data from multiple sources. Cloud platforms provide scalable storage solutions and powerful computing resources to handle large datasets generated by IoT devices. Cloud computing enables researchers, governments, and organizations to access and analyze environmental data remotely, facilitating collaboration and decision-making. In environmental monitoring, cloud platforms can be used to store long-term data on air quality, water levels, or soil conditions, enabling trend analysis, forecasting, and reporting.

- **Hybrid Systems**: The combination of edge and cloud computing provides a hybrid approach that maximizes the strengths of both. Edge devices can handle time-sensitive tasks that require real-time data processing, while cloud systems can perform more complex analyses, handle large-scale data storage, and support Machine Learning models. For example, an IoT network for monitoring water quality may use edge computing to process sensor data and trigger local alerts when contaminants are detected, while sending aggregated data to the cloud for trend analysis, predictive modeling, and long-term monitoring.

- **Improved Scalability and Interoperability**: Advances in both edge and cloud computing allow IoT systems to scale more easily, enabling the deployment of large networks of sensors across diverse environments. This scalability is crucial for monitoring large geographical areas, such as forests, oceans, or urban environments. Furthermore, cloud-based IoT platforms are increasingly designed with interoperability in mind, ensuring that different types of sensors, devices, and systems can seamlessly communicate and work together. This is particularly important in environmental monitoring, where data from multiple sources, such as satellites, drones, and ground-based sensors, must be integrated for comprehensive analysis.

The future of environmental monitoring lies in the continued integration of emerging technologies such as blockchain, autonomous systems, and advanced IoT computing. Blockchain will ensure the integrity and transparency of environmental data, providing a secure framework for tracking environmental impacts and promoting sustainability. Autonomous systems like drones and robots will expand the reach and capabilities of monitoring efforts, enabling real-time data collection in even the most challenging environments. The combination of edge and cloud computing will improve the efficiency, scalability, and real-time processing capabilities of IoT systems, providing powerful tools for addressing pressing environmental challenges. These technological advances will play a crucial role in enabling more effective, efficient, and equitable environmental monitoring, fostering a sustainable future where decision-making is informed by accurate, real-time data and powered by cutting-edge technologies.

16

Policy and Governance

16.1 Global Environmental Monitoring Policies

Environmental monitoring is a critical aspect of global efforts to address climate change, pollution, resource depletion, and biodiversity loss. Across the world, various policies and frameworks have been established to guide environmental monitoring efforts, ensuring that data are collected systematically and used effectively to promote sustainability and public health.

- **International Agreements and Conventions**: Global environmental monitoring is often driven by international treaties and agreements. For example, the United Nations (UN) has established frameworks such as the Paris Agreement on climate change, which requires countries to monitor and report on their greenhouse gas emissions and other environmental indicators. These agreements often rely on global data collection systems, including satellite-based monitoring and ground-level sensors, to track environmental changes and assess the effectiveness of climate mitigation and adaptation measures.

- **Sustainable Development Goals (SDGs)**: The UN's SDGs provide a comprehensive framework for monitoring and managing global environmental sustainability. Goal 13 (Climate Action), Goal 6 (Clean Water and Sanitation), and Goal 15 (Life on Land) highlight the importance of robust environmental monitoring systems. As part of these global efforts, countries are encouraged to improve their environmental data collection capabilities, making use of modern technologies such as Artificial Intelligence (AI), Internet of Things (IoT), and remote sensing to enhance monitoring accuracy, coverage, and timeliness.

DOI: 10.1201/9781003712701-20

- **Regional and National Policies**: In addition to global policies, many countries and regions have their own regulations and initiatives to support environmental monitoring. For example, the European Union (EU) has implemented the European Environmental Agency (EEA), which provides comprehensive data on air and water quality, biodiversity, and climate change. Similarly, countries like the United States have established regulatory bodies like the Environmental Protection Agency (EPA), which enforces environmental monitoring standards and conducts large-scale data collection efforts to track pollution and assess environmental health.

16.2 Regulatory Frameworks for AI and IoT Integration

The integration of AI and IoT in environmental monitoring presents new challenges and opportunities for policymakers. As these technologies become more widespread, it is essential to establish regulatory frameworks that ensure their responsible, ethical, and effective use.

- **Data Protection and Privacy**: With the use of IoT sensors to collect real-time environmental data, concerns about data privacy and security become increasingly important. Regulatory frameworks must establish clear guidelines on the collection, storage, and sharing of data to ensure that individuals' privacy rights are protected. This includes addressing issues such as consent for data collection, ownership of environmental data, and preventing unauthorized access to sensitive information. For example, the EU's General Data Protection Regulation (GDPR) has set a precedent for how data privacy should be handled, including the collection of environmental data through IoT systems.
- **Ethical AI Use**: As AI models are used to analyze environmental data and make predictions about pollution, climate change, and resource management, ethical considerations must be integrated into regulatory frameworks. This includes ensuring that AI algorithms are transparent, fair, and accountable. Regulations must address potential biases in AI models, ensure equitable decision-making, and prevent discriminatory outcomes that disproportionately affect marginalized communities.

- **IoT Device Standards**: For IoT systems to function effectively in environmental monitoring, there needs to be standardization of hardware and software. Regulatory bodies must set technical standards to ensure interoperability between devices and platforms. This includes developing protocols for data transmission, ensuring that sensors are calibrated correctly, and establishing requirements for data accuracy and reliability. For example, the Institute of Electrical and Electronics Engineers has established standards for IoT devices, which could be used as a model for environmental applications.

- **Environmental Impact Assessments**: As AI and IoT technologies are integrated into environmental monitoring, they can have an impact on ecosystems and human health. Regulatory frameworks must require environmental impact assessments for new technologies or systems to evaluate their potential effects before deployment. This ensures that technologies do not inadvertently contribute to environmental degradation or harm vulnerable communities.

16.3 Role of Stakeholders in Policy Implementation

The successful implementation of environmental monitoring policies, particularly those involving AI and IoT technologies, requires the active involvement of multiple stakeholders. These include governments, international organizations, private companies, civil society, local communities, and the scientific community.

- **Governments**: Governments play a central role in shaping environmental policies, funding research, and establishing regulatory frameworks. They are responsible for enforcing laws and regulations that promote sustainable environmental monitoring practices. Governments must also collaborate with international bodies to ensure that global standards for environmental data collection and analysis are met. Additionally, they can incentivize the adoption of AI and IoT technologies in environmental monitoring through grants, tax breaks, and other support mechanisms.

- **Private Sector**: Private companies, particularly those in the technology, manufacturing, and telecommunications sectors, are key contributors to the development and deployment of AI and IoT solutions for environmental monitoring. These companies are responsible for creating the sensors, devices, software, and communication networks that power modern environmental monitoring systems. Collaboration between the public and private sectors is essential to ensure that technologies are designed to meet the needs of environmental monitoring while also adhering to regulatory standards.

- **Scientific Community**: Researchers and scientists play an essential role in developing the methodologies, technologies, and analytical techniques that support environmental monitoring efforts. Their expertise is critical in designing experiments, interpreting data, and advancing new technologies that can improve the accuracy and effectiveness of environmental monitoring. Moreover, scientists provide evidence-based recommendations to policymakers to guide decision-making.

- **Local Communities**: The involvement of local communities is crucial for the success of environmental monitoring policies, particularly in rural or marginalized areas. Local communities often possess valuable traditional knowledge and insights into the local environment that can complement technological data. Engaging communities in data collection efforts, policy implementation, and monitoring helps ensure that policies are context-specific, culturally appropriate, and address local needs. It also fosters greater community ownership and trust in environmental monitoring systems.

- **Non-Governmental Organizations (NGOs) and Civil Society**: NGOs and civil society organizations often act as advocates for environmental protection, ensuring that policies are equitable and inclusive. These organizations can play a key role in raising awareness about environmental issues, conducting independent monitoring, and holding governments and corporations accountable for their environmental actions. They also help bridge the gap between scientific research, policy, and public participation.

Effective environmental monitoring policies are essential for addressing global challenges such as climate change, biodiversity loss, and pollution. As AI and IoT technologies become more integral to environmental monitoring systems, it is vital to establish strong regulatory frameworks that promote data integrity, ethical decision-making, and data security. Governments, the private sector, the scientific community, and civil society all play critical roles in implementing these policies and ensuring that environmental monitoring efforts are transparent, equitable, and effective. By working together, stakeholders can harness the power of technology to promote a sustainable and resilient future.

17

SOME BENEFITS AND TIPS

1. Why Host Online Materials?

- **Practical Experience**: Readers gain immediate access to datasets and tutorials, avoiding technical setup issues.
- **Current Materials**: Quickly refresh datasets and tutorials as new information or techniques become available.
- **Fostering Community**: Create spaces for readers to exchange ideas, seek help, and collaborate via comments or forums.
- **Rich Media**: Incorporate videos, interactive notebooks (e.g., Jupyter), and web applications to enhance comprehension.
- **Inclusive Access**: Reduce obstacles for users with limited resources by providing free software and data access.

2. What to Host Online?

- **Sample datasets** (with links and instructions)
- **Step-by-step tutorial code** and notebooks (Python, R, etc.)
- **Demo applications or web tools** (like image classifiers or acoustic analyzers)
- **Additional reading and resources** (papers, blogs, videos)
- **Community forum or discussion board** for readers

3. Tools and Platforms for Hosting

- **GitHub or GitHub Pages**: For code, datasets, and static site hosting
- **Google Colab**: Share runnable notebooks for free cloud execution
- **Hugging Face Spaces**: Host AI demos and models interactively
- **YouTube or Vimeo**: For tutorial videos
- **Discourse or Reddit**: For community discussions

18

CHALLENGES IN AI ENVIRONMENTAL PROJECTS

1. Data-Related Challenges

- **Limited Availability and Data Quality**: Many marginalized areas suffer from scarce or outdated environmental datasets. Satellite imagery might be expensive or low in detail, while terrestrial sensors can be few or absent.

    Approach: Utilize citizen science contributions, apply data augmentation methods, and employ transfer learning to maximize limited data resources.

- **Bias and Lack of Representativeness**: AI systems trained on datasets from specific regions often struggle to perform well elsewhere due to differences in ecology, culture, or socioeconomic factors.

    Approach: Implement inclusive data gathering and apply techniques to reduce model bias.

- **Privacy and Data Sovereignty Concerns**: Indigenous and local communities may worry about control over their data and potential misuse.

    Approach: Ensure transparent data governance, foster partnerships with local groups, and establish data-sharing protocols that honor regional laws and customs.

2. Infrastructure and Technical Constraints

- **Connectivity and Power Constraints**: Remote and underserved regions often face unstable internet access and limited electricity supply.

    Solution: Deploy edge computing, use solar-powered Internet of Things (IoT) devices, and develop Artificial Intelligence (AI) tools that work offline.

DOI: 10.1201/9781003712701-22

- **High Computational Demands**: Training complex AI models requires GPUs and cloud infrastructure that may be costly or unavailable locally.

  Solution: Utilize cloud funding programs like Microsoft AI for Earth, apply model compression techniques, and adopt federated learning approaches.

3. **Human Capacity and Governance**
   - **Shortage of Local AI Skills**: It can be challenging to develop and keep data science and AI experts within communities, causing dependence on outside specialists.

     Approach: Invest in training initiatives, provide remote mentoring, and collaborate closely with local partners.
   - **Regulatory and Policy Gaps**: Numerous nations do not have established rules governing AI applications in environmental sectors, which may lead to misuse or public skepticism.

     Approach: Promote the creation of AI governance policies emphasizing openness, equity, and long-term sustainability.

19

DIGITAL TWIN TECHNOLOGY IN ENVIRONMENTAL APPLICATIONS

19.1 What Is a Digital Twin?

A digital twin is a real-time virtual replica of a physical system created through sensor inputs, simulations, and Artificial Intelligence. It allows for monitoring, experimentation, and prediction without disturbing the real-world environment. Applications in environmental studies include

- **Urban Ecosystems**:
  Modeling air pollution, vehicle emissions, and the design of urban green areas.
- **Water Resource Management**:
  Virtual replicas of river basins or watersheds used to improve irrigation efficiency, forecast flooding, and oversee drought conditions.
- **Forest and Wildlife Habitats**:
  Simulating tree development, wildlife movement, and wildfire hazards.
- **Technologies Behind Digital Twins**
  - Incorporating Internet of Things sensors to collect live data streams
  - Using geospatial information and geographic information system layers to provide spatial awareness
  - Applying AI techniques for forecasting and identifying irregularities
  - Leveraging cloud services for storing, analyzing, and displaying data

DOI: 10.1201/9781003712701-23

20

EXPANDED SECTION

GIS and Remote Sensing Essentials

1. **Geographic Information Systems (GIS)**
   - **Fundamentals**: Data layers, differences between vector and raster formats, and coordinate reference systems
   - **Applications in Environment**: Mapping habitats, planning land use, monitoring deforestation and urban expansion
   - **Common Software**:
     - QGIS (open-source, popular especially in developing regions)
     - ArcGIS (commercial software with advanced analytics and enterprise capabilities)
     - Google Earth Engine (cloud platform offering extensive satellite data and powerful processing)
2. **Remote Sensing Fundamentals**
   - **Data Sources**: Satellites such as Sentinel and Landsat, unmanned aerial vehicles (drones), and aerial imaging
   - **Data Formats**: Multispectral imagery (various color bands), hyperspectral data (numerous narrow spectral bands), and LiDAR-generated 3D point clouds
   - **Use Cases in Environment**:
     - Assessing plant health and agricultural productivity
     - Identifying unauthorized mining or deforestation activities
     - Charting flood zones and areas affected by wildfires

DOI: 10.1201/9781003712701-24

21

UNCERTAINTY ANALYSIS IN AI FOR ENVIRONMENTAL RESEARCH

21.1 Definition

Uncertainty analysis involves evaluating and measuring doubt in data, models, and predictions. Acknowledging uncertainty is vital for trustworthy environmental decision-making amid complex systems.

Sources of Uncertainty:

- **Data issues**: Errors in measurement, gaps, and coarse resolution.
- **Model Limitations**: Simplifications, assumptions, and insufficient training data.
- **Environmental Dynamics**: Natural variability and unpredictable phenomena.
- **Human Influence**: Subjective decisions in data processing and modeling choices.

Why It's Crucial:

- **Risk Management**: Quantifies confidence levels in forecasts (e.g., floods).
- **Clear Communication**: Transparency fosters stakeholder confidence.
- **Better Decisions**: Supports adaptive strategies and contingency plans.
- **Model Refinement**: Highlights weak points and informs future data collection.

DOI: 10.1201/9781003712701-25

Methods for Quantifying Uncertainty:

- **Probabilistic Approaches**: Bayesian inference, Monte Carlo simulations.
- **Ensemble Techniques**: Combining multiple model outcomes to evaluate variability.
- **Sensitivity Testing**: Examining how input changes impact results.
- **Error Tracking**: Following uncertainty through calculation steps.

Advice for Practitioners:

- Always report uncertainty alongside predictions.
- Use visuals like confidence intervals or heatmaps to illustrate uncertainty.
- Involve stakeholders in interpreting uncertain outcomes.
- Utilize uncertainty insights to focus data gathering and model improvements.

22

PRACTICAL HANDS-ON
TUTORIALS (EXAMPLES)

- **Simple Land Cover Classification Using Sentinel-2 in Google Earth Engine**

  A stepwise tutorial on loading satellite imagery, choosing training areas, and executing a classification algorithm to distinguish forests from agricultural land.

- **Basic Digital Twin Development for a River Basin with IoT Data and Python**

  Gathering sensor readings for water levels, temperature, and precipitation, creating live visual dashboards, and modeling flood events.

- **Acoustic Species Detection Using Open Sound Datasets**

  Building a convolutional neural network to identify bird calls from remote audio recordings, determining species presence or absence.

22.1 Additional Practical Tips and Resources

- **Utilizing Open-Source and Free Software**:

  Focus on tools like QGIS, Google Earth Engine, and Python frameworks such as TensorFlow, PyTorch, and Rasterio.

- **Forming Local Collaborations**:

  Work together with academic institutions, NGOs, and community groups to gain data access, maintain project relevance, and foster trust.

- **Progressive Development**

  Begin with basic prototypes or pilot studies, then expand and improve based on ongoing feedback and evaluation.

DOI: 10.1201/9781003712701-26

23
CONCLUDING PERSPECTIVES

23.1 Summary of Key Concepts

The integration of Artificial Intelligence (AI) and the Internet of Things (IoT) in environmental monitoring marks a pivotal advancement in the way we track, analyze, and manage environmental data. AI brings powerful tools such as Machine Learning and predictive modeling, which enable the interpretation of vast datasets to uncover trends, identify patterns, and forecast future environmental conditions. IoT, on the other hand, allows for the widespread deployment of sensors that gather real-time data across diverse environments, facilitating continuous and localized monitoring.

Key environmental parameters such as air quality, water quality, soil health, and climate variables are now being more accurately monitored and managed through AI and IoT systems. These systems can be applied in various domains, including air pollution control, water contamination detection, disaster management, and agricultural sustainability. Furthermore, the convergence of these technologies offers new possibilities for environmental governance, policy implementation, and global cooperation in tackling climate change and promoting sustainable resource management.

23.2 Challenges and Opportunities in AI and IoT Applications

While the potential of AI and IoT in environmental monitoring is immense, several challenges need to be addressed to fully harness their capabilities.

- **Data Accuracy and Quality**: The effectiveness of AI and IoT systems is heavily dependent on the quality of the data they process. Inaccurate, incomplete, or biased data can lead to erroneous conclusions and ineffective policy recommendations.

Ensuring that sensors are properly calibrated and that data are consistent and reliable is essential. Advances in sensor technology and AI algorithms, such as data cleaning techniques and robust Machine Learning models, are helping mitigate these issues, but challenges remain.

- **Data Privacy and Security**: The use of IoT devices in environmental monitoring raises concerns about data privacy and security. Given the scale at which data are collected, ensuring that sensitive data are protected from cyber threats and unauthorized access is paramount. Furthermore, issues of data ownership and sharing must be carefully considered, particularly in collaborative efforts between governments, private companies, and local communities. Transparent data governance models and secure communication protocols will be crucial to overcome these challenges.

- **Infrastructure and Connectivity**: Deploying large-scale IoT networks in remote or underdeveloped regions remains a significant challenge. Reliable network infrastructure, particularly in rural or underserved areas, is often lacking, making it difficult to deploy IoT systems that can collect and transmit data in real time. However, the emergence of low-power wide-area networks, 5G technology, and satellite-based communication systems offers opportunities to overcome these barriers and expand IoT applications in environmental monitoring.

- **Ethical and Social Implications**: The use of AI and IoT in environmental monitoring brings ethical considerations regarding decision-making, equity, and inclusivity. Ensuring that these technologies are used to promote social justice, minimize bias, and involve affected communities in the monitoring process is critical. Moreover, regulatory frameworks must address the ethical use of AI, ensuring that algorithms do not inadvertently harm vulnerable populations or the environment.

On the other hand, these challenges present numerous opportunities for innovation and growth. Advancements in AI algorithms, edge computing, and IoT hardware will continue to improve the accuracy, scalability, and sustainability of environmental monitoring systems. Moreover, increased global cooperation and the development of common standards for environmental data sharing will pave the way for more effective cross-border environmental management.

23.3 Vision for the Future of Environmental Monitoring

The future of environmental monitoring lies in the seamless integration of AI, IoT, and other emerging technologies to create smart systems capable of delivering real-time, actionable insights for environmental management. With the rapid growth of sensor networks, improved AI algorithms, and greater computational power, the ability to monitor the environment at an unprecedented scale and detail is within reach.

1. **Predictive Environmental Management**: In the future, environmental monitoring systems will not only detect and report environmental parameters but also predict trends and outcomes. AI models will leverage historical and real-time data to forecast air pollution levels, water quality changes, and climate events, allowing for proactive management and mitigation strategies. This predictive capability could extend to disaster preparedness, enabling early warning systems for events such as floods, wildfires, and hurricanes, minimizing loss of life and property.

2. **Decentralized and Autonomous Systems**: The growing integration of IoT devices and AI will result in decentralized monitoring networks, where sensors and autonomous systems (such as drones, robots, and AI-powered devices) collect and analyze data in real time, making decisions without the need for human intervention. This will significantly enhance the efficiency of environmental monitoring, especially in remote or hazardous environments. Autonomous systems will play a key role in environmental cleanup efforts, wildlife monitoring, and land management, improving the speed and precision of interventions.

3. **Global Environmental Cooperation and Transparency**: As data-sharing platforms become more sophisticated, countries and organizations will have greater access to real-time

environmental data, fostering global collaboration in addressing cross-border environmental issues. Transparency in environmental data will allow for greater accountability in policy implementation and monitoring. Blockchain technology, integrated with IoT systems, could further enhance trust by ensuring the integrity of environmental data, making it accessible and verifiable by all stakeholders, including governments, corporations, and the public.

4. **Smart Cities and Sustainable Development**: The future of environmental monitoring is closely tied to the development of smart cities. IoT and AI technologies will be integrated into urban planning and infrastructure, enabling smart grids, waste management systems, and transportation networks that are more sustainable and efficient. By integrating environmental monitoring systems into urban landscapes, cities will be able to manage resources more effectively, reduce pollution, and improve the quality of life for residents.

5. **Empowering Local Communities**: AI and IoT systems will empower local communities by providing them with tools to monitor and manage their own environments. Mobile apps, low-cost sensors, and community-based data collection initiatives will enable citizens to play an active role in environmental protection. This democratization of environmental data can lead to greater public awareness, engagement, and advocacy, helping to address environmental challenges at the grassroots level.

The convergence of AI, IoT, and environmental monitoring technologies promises a future where the environment is managed more efficiently, sustainably, and equitably. However, to realize this vision, a concerted effort is needed to address the challenges of data accuracy, privacy, infrastructure, and ethical concerns. By fostering innovation, collaboration, and thoughtful policy development, we can build a world where environmental monitoring systems not only protect our planet but also empower individuals and communities to take action for a sustainable future. The future of environmental monitoring is not just about collecting data, but using that data to create a smarter, more resilient world for generations to come.

References

Abbas, Z., & Yoon, W. (2015). A survey on energy conserving mechanisms for the internet of things: Wireless networking aspects. *Sensors*, 15(10), 24818–24847.

Ahbil, K., Sellami, F., Baati, H., Gautam, S., & Azri, C. (2024). Influence of localized sources and meteorological conditions on dry-deposited particles: A case study of Gabès, Tunisia. *Science of the Total Environment*, 954, 176726.

Ahmad, S., & Ahmad, T. (2023). AQI prediction using layer recurrent neural network model: A new approach. *Environmental Monitoring and Assessment*, 195(10). https://doi.org/10.1007/s10661-023-11646-3

Alaba, F. A., Othman, M., Hashem, I. A. T., & Alotaibi, F. (2017). Internet of Things security: A survey. *Journal of Network and Computer Applications*, 88, 10–28. https://doi.org/10.1016/j.jnca.2017.04.002

Alam, M. G. (2017). CNN based mood mining through IoT-based physiological sensors observation. *Journal of the Korean Data & Information Science Society*, 2017, 1301–1303.

Alzubaidi, L., Zhang, J., Humaidi, A. J., Al-Dujaili, A., Duan, Y., Al-Shamma, O., Santamaría, J., Fadhel, M. A., Al-Amidie, M., & Farhan, L. (2021). Review of deep learning: Concepts, CNN architectures, challenges, applications, future directions. *Journal of Big Data*, 8, 53. https://doi.org/10.1186/s40537-021-00444-8

Ambade, B., Kumar, T. K., Gautam, S., Mahto, D. K., Dumka, U. C., Mohammad, F., Al-Lohedan, H. A., Soleiman, A. A., & Gautam, A. S. (2023). Black carbon vs carbon monoxide: Assessing the impact on Indian urban cities. *Water, Air, & Soil Pollution*. https://doi.org/10.1007/s11270-023-06706-w

Asha, P., Natrayan, L., Geetha, B. T., Beulah, J. R., Sumathy, R., Varalakshmi, G., & Neelakandan, S. (2022). IoT enabled environmental toxicology for air pollution monitoring using AI techniques. *Environmental Research*, 205, Article 112574.

Azri, C. (2000). Contribution des sources mobiles et fixes à la pollution atmosphérique dans la région de Sfax (Tunisie). Thèse de Doctorat, Université de Tunis II, Tunisie.

Azri, C., Abida, H., & Medhioub, K. (2009). Geochemical behavior of the Tunisian background aerosols in sirocco wind circulations. *Advances in Atmospheric Sciences*, 26(3), 390–402.

Azri, C., Abida, H., & Medhioub, K. (2010). Geochemical behaviour of the aerosol sampled in a suburban zone of Sfax City (Tunisia). *International Journal of Environment and Pollution*, 41(1–2), 51–69.

Baati, H., Siala, M., Azri, C., Ammar, E., Dunlap, C., & Trigui, M. (2020). Resistance of a *Halobacterium salinarum* isolate from a solar saltern to cadmium, lead, nickel, zinc, and copper. *Antonie Van Leeuwenhoek*, 113, 1699–1711.

Bahloul, M., Chabbi, I., Sdiri, A., Amdouni, R., Medhioub, K., & Azri, C. (2015). Spatiotemporal variation of particulate fallout instances in Sfax City, Southern Tunisia: Influence of sources and meteorology. *Advances in Meteorology*, 2015(11), Article ID 471396. https://doi.org/10.1155/2015/471396

Blessy, A., Paul, J. J., Gautam, S., Shany, V. J., & Sreenath, M. (2023). IoT-based air quality monitoring in hair salons: Screening of hazardous air pollutants based on personal exposure and health risk assessment. *Water, Air, & Soil Pollution*. https://doi.org/10.1007/s11270-023-06350-4

Blumenstock, J. (2020). Machine learning can help get covid-19 aid to those who need it most. *Nature*. https://doi.org/10.1038/d41586-020-01393-7. Epub ahead of print. PMID: 32409767.

Broy, M., Cengarle, M. V., & Geisberger, E. (2020). Cyber-physical systems: Imminent challenges. *ACM Transactions on Cyber-Physical Systems*, 8(2), 47. https://doi.org/10.1007/978-3-642-34059-8_1

Buelvas, J., Múnera, D., Tobón, D. P. V., Aguirre, J., & Gaviria, N. (2023). Data quality in IoT-based air quality monitoring systems: A systematic mapping study. *Water, Air, & Soil Pollution*, 234(4), 248.

Chabbi, I., Bahloul, M., Dammak, R., & Azri, C. (2017). Dust particles deposition quality assessment in rural areas located not far from a congested highway and several sebkhas: Case of Monastir region, eastern Tunisia. *Journal of Environmental Engineering and Landscape Management*, 26(2), 141–157.

Chabbi, I., Baati, H., Dammak, R., Bahloul, M., & Azri, C. (2021). Toxic metal pollution and ecological risk assessment in superficial soils of "rural-agricultural and coastal-urban" of Monastir region, Eastern Tunisia. *Human and Ecological Risk Assessment*, 27(3), 375–594. https://doi.org/10.1080/10807039.2020.1732189

Campbell, S. J., Wolfer, K., Utinger, B., Westwood, J., Zhang, Z. H., Bukowiecki, N., Steimer, S. S., Vu, T. V., Xu, J. S., Straw, N., Thomson, S., Elzein, A., Sun, Y. L., Liu, D., Li, L. J., Fu, P. Q., Lewis, A. C., Harrison, R. M., Bloss, W. J., Loh, M., Miller, M. R., Shi, Z. B., & Kalberer, M. (2021). Atmospheric conditions and composition that influence PM2.5 oxidative potential in Beijing, China. *Atmospheric Chemistry and Physics*, 21, 5549–5573. https://doi.org/10.5194/acp-21-5549-2021

Chen, L., Wang, Y., & Zhang, H. (2016). IoT for forest fire monitoring: Integrating sensor networks with satellite data. *International Journal of Wildland Fire*, 15(3), 321–335.

Copernicus Climate Change Service. (2023). Real-time monitoring tools. Retrieved from https://climate.copernicus.eu

Cui, S., Gao, Y., Huang, Y., Shen, L., Zhao, Q., Pan, Y., & Zhuang, S. (2023). Advances and applications of machine learning and deep learning in environmental ecology and health. *Environmental Pollution*, 335, 122358. https://doi.org/10.1016/J.ENVPOL.2023.122358

Dammak, R., Bahloul, M., Chabbi, I., & Azri, C. (2016). Spatial and temporal variations of dust particle deposition at three urban/suburban areas in Sfax city (Tunisia). *Environmental Monitoring and Assessment*, 188, 336. https://doi.org/10.1007/s10661-016-5341-0

Dammak, R., Chabbi, I., Bahloul, M., & Azri, C. (2020). PM10 temporal variation and multi-scale contributions of sources and meteorology in Sfax, Tunisia. *Air Quality, Atmosphere, and Health*, 13, 617–628. https://doi.org/10.1007/s11869-020-00824-8

Dhyani, M, & Kumar, R. (2021). An intelligent Chatbot using deep learning with bidirectional RNN and attention model. *Materials Today: Proceedings*, 34, 817–824.

Du, Y., Ma, C., Wu, C., Xu, X., Guo, Y., Zhou, Y., & Li, J. (2017). A visual analytics approach for station-based air quality data. *Sensors (Switzerland)*, 17(1), 1–17. https://doi.org/10.3390/s17010030

Dua, S., & Du, X. (2011). Data mining and machine learning in cybersecurity. CRC press, Taylor & Francis Group.

Dunster, C., Oliete, A., Jacob, V., Besombes, J. L., Chevrier, F., & Jaffrezo, J. L. (2018). Comparison between five acellular oxidative potential measurement assays performed with detailed chemistry on PM10 samples from the city of Chamonix (France). *Atmospheric Chemistry and Physics*, 18, 7863–7875. https://doi.org/10.5194/acp-18-7863-2018

Gupta, A., Bherwani, H., Gautam, S., Anjum, S., Musugu, K., Kumar, N., Anshul, A., & Kumar, R. (2021). Air pollution aggravating COVID-19 lethality? Exploration in Asian cities using statistical models. *Environment, Development and Sustainability*, 23, 6408–6417.

Gautam, S., Kumar, A. P., Prakash, S. S., & Hitch, M. (2016). Particulate matter pollution in opencast coal mining areas: A threat to human health and environment. *International Journal of Mining, Reclamation and Environment*, 32(2), 75–92.

Gope, P., & Hwang, T. (2016). BSN-care: A secure IoT-based modern health-care system using body sensor network. *IEEE Sensors Journal*, 16(5), 1368–1376. https://doi.org/10.1109/JSEN.2015.2502401

Gopichand, G., Sarath, T., Dumka, A., Goyal, H. R., Singh, R., Gehlot, A., Gupta, L. R., Thakur, A. K., Priyadarshi, N., & Twala, B. (2024). Use of IoT sensor devices for efficient management of healthcare systems: A review. *Discover Internet of Things*, 4, 8. https://doi.org/10.1007/s43926-024-00062-9

Ham, Y. G., Kim, J. H., & Luo, J. J. (2019). Deep learning for multi-year ENSO forecasts. *Nature Communications*, 10(1), 1–7. https://doi.org/10.1038/s41467-019-09699-2

Hussain, J. A., Ambade, B., Sankar, T. K., Mohammad, A. A., Soleiman, F., & Gautam, S. (2023). Black carbon emissions in the rural Indian households: Sources, exposure, and associated threats. *Geological Journal*, https://doi.org/10.1002/gj.4775

Infocomm Media Development Authority. (2020). Singapore's model AI governance framework. Retrieved from https://imda.gov.sg

Jeong, Y., Schäfer, A., & Smith, K. (2018). A comparison of equilibrium and kinetic passive sampling for the monitoring of aquatic organic contaminants in German rivers. *Water Research*, 145, 248–258. https://doi.org/10.1016/j.watres.2018.08.016

Jobin, A., Ienca, M., & Vayena, E. (2019). Artificial intelligence: The global landscape of ethics guidelines. *Nature Machine Intelligence*, 1(9), 389–399. https://doi.org/10.1038/s42256-019-0088-2

Jones, L., Ronan, J., McHugh, B., & Regan, F. (2019). Passive sampling of polar emerging contaminants in Irish catchments. *Water Science and Technology*, 79, 218–230. https://doi.org/10.2166/wst.2019.021

Kaserzon, S. L., Hawker, D. W., & Booij, K. (2014). Passive sampling of perfluorinated chemicals in water: In-situ calibration. *Environmental Pollution*, 186, 98–103. https://doi.org/10.1016/j.envpo l.2013.11.030

Kim, D., Han, H., Wang, W., Kang, Y., Lee, H., & Kim, H. S. (2022). Application of deep learning models and network method for comprehensive air-quality index prediction. *Applied Sciences (Switzerland)*, 12(13). https://doi.org/10.3390/app12136699

Li, Y., Yao, C., Zha, D., Yang, W., & Lu, G. (2018). Selection of performance reference compound (PRC) for passive sampling of pharmaceutical residues in an effluent dominated river. *Chemosphere*, 211, 884–892. https://doi.org/10.1016/j.chemosphere.2018.07.179

Liu, X., Lu, D., Zhang, A., Liu, Q., & Jiang, G. (2022). Data-driven machine learning in environmental pollution: Gains and problems. *Environmental Science & Technology*, 56, 2124–2133.

Majid Butt, M. et al. (2023). Ambient IoT: A missing link in 3GPP IoT Devices Landscape. Retrieved from https://arxiv.org/html/2312.06569v1. License: CC BY 4.0, arXiv:2312.06569v1.

Minh, V. T. T., Tin, T. T., & Hien, T. T. (2021). PM2.5 Forecast System by Using Machine Learning and WRF model, a case study: Ho Chi Minh City, Vietnam. *Aerosol and Air Quality Research*, 21(12). https://doi.org/10.4209/AAQR.210108

Mkawar, S., Azri, C., Kamoun, F., & Montacer, M. (2007). Impact sur les biophases marines des rejets anthropiques, notamment des métaux lourds rejetés sur le littoral nord de la ville de Sfax (Tunisie). *Techniques – Sciences – Methodes*, 10, 71

Mumtaz, R., Zaidi, S. M. H., Shakir, M. Z., Shafi, U., Malik, M. M., Haque, A., & Zaidi, S. A. R. (2021). Internet of things (Iot) based indoor air quality sensing and predictive analytic—A COVID-19 perspective. *Electronics*, 10(2), 184.

NASA Cryosphere Program. (2023). AI applications in Arctic ice monitoring. Retrieved from https://www.nasa.gov/cryosphere

NIST. (2021). Cybersecurity framework for IoT devices. Retrieved from https://www.nist.gov

Orru, H., Ebi, K., & Forsberg, B. (2017). The interplay of climate change and air pollution on health. *Current Environmental Health Reports*, 4, 504–513.

Packiavathy, S. V., & Gautam, S. (2023). Internet of Things (IoT) based automated sanitizer dispenser and COVID-19 statistics reporter in a post-pandemic world. *Health and Technology*, 13(2), 327–341.

Pan, W. (2016). A survey of transfer learning for collaborative recommendation with auxiliary data. *Neurocomputing*, 177, 447–453.

Ramya, K., Nargees, S., Tabasuum, S. A., Khan, S., & Shin, M. A. (2020). Survey onSmart automated WheelChair system with voice controller using IOT along with health monitoring for physically challenged persons. *International Journal of Management Science and Engineering Management*, 5, 95–98.

Rasp, S., Pritchard, M., & Gentine, P. (2020). WeatherBench: A benchmark dataset for data-driven weather forecasting. *Journal of Advances in Modeling Earth Systems*, 12(11), e2020MS002203. https://doi.org/10.1029/2020MS002203

Rustam, F., Ishaq, A., Kokab, S. T., de la Torre Diez, I., Mazón, J. L. V., Rodríguez, C. L., & Ashraf, I. (2022). An artificial neural network model for water quality and water consumption prediction. *Water (Switzerland)*, 14(21), 1–23. https://doi.org/10.3390/w14213359

Sarker, I. H. (2022). AI-based modeling: Techniques, applications and research issues towards automation, intelligent and smart systems. *SN Computer Science*, 3, 158. https://doi.org/10.1007/s42979-022-01043-x

Selvadass, S., Paul, J. J., Mary, I. T. B., Packiavathy, I. S. V., & Gautam, S. (2022). IoT-enabled smart mask to detect COVID19 outbreak. *Health and Technology*, 12, 1025–1036.

Sellami, F., Dammak, R., & Azri, C. (2022). Analysis of daily and diurnal O_3-NOx relationships and assessment of local/regional oxidant $(OX = O_3 + NO_2)$ levels and associated human health risk at a coastal suburban site of Sfax (Tunisia). *Archives of Environmental Contamination and Toxicology*, 84(1), 119–136. https://doi.org/10.1007/s00244-022-00966-z

Sellami, F., Baati, H., & Azri, C. (2023). Airborne particulates pollution level characterization and ecological health risk assessment of toxic metals in suburban Sfax, Southern Tunisia. *Water, Air, and Soil Pollution*, 234, 553. https://doi.org/10.1007/s11270-023-06581-5

Sellami, F., Mazzi, G., Gautam, S., Gambaro, A., & Azri, C. (2025). Assessment of air pollution and associated health risks in Central Tunisia (North Africa) during and outside Saharan dust events. *Stochastic Environmental Research and Risk Assessment*. https://doi.org/10.1007/s00477-025-03017-w

Shorten, C., & Khoshgoftaar, T. M. (2019). A survey on image data augmentation for deep learning. *Journal of Big Data*, 6(1), 60.

Tastan, M. (2018). IoT based wearable smart health monitoring system. *Celal Bayar University Journal of Science*, https://doi.org/10.18466/cbayarfbe.451076

Urban, M. C., Bocedi, G., Hendry, A. P., Mihoub, J. B., Pe'er, G., Singer, A., & Travis, J. M. J. (2020). Improving biodiversity projections under climate change. *Science*, 362(6412), 292–293. https://doi.org/10.1126/science.aas4868

Wang, F., Wang, H., Wang, H., Li, G., & Situ, G. (2019). Learning from simulation: An end-to-end deep-learning approach for computational ghost imaging. *Optics Express*, 27(18), 25560–25572.

Weiss, K., Khoshgoftaar, T. M., & Wang, D. (2016). A survey of transfer learning. *Journal of Big Data*, 3(1), 9.

Xu, Z., Li, D., & Liu, J. (2019). Evolution of environmental monitoring methods: A comprehensive review. *Environmental Science & Technology*, 53(4), 1712–1734. https://doi.org/10.1021/es900933u

Zaïbi, C., Carbonel, P., Kamoun, F., Fontugne, M., Azri, C., Jedoui, Y., & Montacer, M. (2012). Évolution de la sebkha Dreîaa (Sud-Est de la Tunisie, Golfe de Gabès) au cours de l'Holocène supérieur: réponse des associations des ostracodes. *Revue de Micropaleontologie*, 55(3), 83–97.

Index

For Product Safety Concerns and Information please contact our EU
representative GPSR@taylorandfrancis.com
Taylor & Francis Verlag GmbH, Kaufingerstraße 24, 80331 München, Germany